Root Cause Data Collection
Using the Arduino

Paul Bradt and David Bradt

Copyright © 2017 P & D Analytics Incorporated

All rights reserved.

ISBN-13: 978-1975887810
ISBN-10: 1975887816

DEDICATION

The authors dedicate this book to all of the Science Technology Engineering Math (STEM) teachers who have guided and shaped the paths of many young minds (including ours) to question, learn, and utilize new technology to solve problems. Without these unsung heroes the world would not have powerful cell phones, highly reliable cars, the internet, and many other amazing things we routinely take for granted.

Acknowledgements i

Preface iii

1. Introduction 5

2. Business Terminology 9

3. Basics of Root Cause Analysis 15

4. Measurement, Arduino UNO, Data Capture/Analysis, and Safety 23

5. Measuring Counts and Timing of Operations 35

6. Measuring Temperature 55

7. Measuring Light Intensity and Temperature 83

8. Measuring Pressure Change 97

9. Root Cause Analysis of Measured Data 105

10. Enclosures 107

Conclusion 115

Appendix 117

About P & D Analytics Inc. 139

Acknowledgements

The authors at P & D Analytics Inc. thank Jared Brank and Dennis Pate for their assistance with the Arduino. Getting the authors started was an immense help. The authors thank Laura Brank and Andrew Bradt for providing additional technical insight and elements to this book. Excellent editorial reviews were provided by Joanna Opaskar and Ed Weisblatt. Most important was the support and advice from Andrea Bradt.

Preface

Assisting small businesses by improving their *root cause analysis skills* characterizes the authors' passion. To achieve this goal, they rely on more than 30 years of analytical experience and 5 years' experience with modern, low-cost microcontroller tools. Their background and tools provide the basis for developing unique, innovative solutions and techniques to gather relevant data and pinpoint the causes of many technical issues.

Effective *root cause analysis* and the resulting solutions reduce rework and rejected products, and almost *immediately improve bottom-line costs*. This analysis also assists in identifying potentially catastrophic issues caused by failures and errors. This can prevent them from occurring or repeating.

This book is one example of how our company can assist you. It provides some of the basic tools to measure and capture data to identify where failures or product defects occur. We encourage your team to use it to develop your own unique solutions.

1. Introduction

This book focuses on utilizing the powerful Arduino micro-controller to gather data for root cause analysis. Useful tools and techniques are presented to gather data and analyze problems in a process or during operations. The principles and tools may apply to a broad variety of conditions, processes, and problems that range from the assembly of a large piece of equipment, to the effects of temperature on drying paint, to operations in the field or in an office environment. Key measurements and data highlighted in this book include: the time it takes to complete a step, temperature changes, and the heating effect of light, among other significant physical factors. Factors such as these in a system or process may have significant impacts along with unintended consequences. Guidance and ideas presented in this book on how to develop measuring and data logging systems can assist in gaining analytical or practical knowledge of these physical traits.

In some cases data already exists to determine how these factors may influence a system and cause lost production or defective products. Sometimes the data are not ready to be collected, difficult to obtain, or do not exist. Before an expensive project is undertaken to implement a complex measurement system, it makes sense to try out a simple and less costly system first.

How to use this book

This book highlights and explains an incredibly useful, simple, low-cost tool that can aid in root cause analysis. The original Arduino Micro Processor Board, developed in 2004, provides a great interface between measurements of interest, the computer, and ultimately the engineer or technical person in charge of the process. It was developed as a tool to teach the basics of programming. Three things make it stand out as a tool of extensive value:

1. The processor is a basic device and relatively simple to program.//
2. A wide range of sensors are available that can interface with the processor.
3. Many on-line resources: including open source code and helpful forums. Additionally, there are a lot of user groups in the US.

The authors recommend that the investigator read through the first introductory chapters (1 to 4). Then, read the descriptions for the project chapters. This will provide a basic understanding of the systems and the questions they can answer. Some chapters build on previous work, so reviewing earlier chapters can aid in understanding how the systems work in later chapters.

For the most part the projects shown are set up as prototypes and not final assemblies. Chapter 10 highlights a few of these projects assembled in enclosures providing protection to the Arduino and appear finished or complete. The investigator has the option to determine whether or not to implement an enclosure for their project.

Even though the first few programs and projects appear relatively simple, they may be the solution needed. They may also provide some insight into other methods and techniques that the investigator may want to utilize to solve their own unique challenge.

The Appendix contains additional helpful information related to finding components, substituting other sensors, soldering safety, assembling the components used in this book, and additional technical information related to the example problem. This information may answer questions the investigator has while reading this book.

Very Helpful Tip related to Programming the Arduino

The authors have discovered an important step that aids in the determination of the version of the code loaded on the Arduino:

1. Name and label the Arduino (in this book it starts out as SN001 and increments from there).

2. When naming the code, use the unique Arduino number, a brief description of the code, and the date. For example, the first project is on the SN001 Arduino, and it captures counts only. So the name used is: "SN001_Counter_only_apr_29"

This system, while not foolproof, allows easy identification of the code loaded on an Arduino for troubleshooting or finding the one needed for a project. This can simplify the configuration management of the Arduino.

Disclaimer: The phrase "Knowledge is Power" describes what these tools may provide. At the same time that power should be used wisely. There are no implied warranties from using these systems. The process owners will use their own experience and analysis of all the data to reach their own decisions.

In root cause analysis, it is wise to focus on systems and processes that support employees rather than blaming them for failures. In most cases they are trying to do their job to the best of their ability given their knowledge, equipment and support.

Ultimately, the authors' goal is to inspire readers to use these ideas to develop their own unique systems and solutions. The Arduino is a powerful, low-cost tool which can be a cost effective and timely solution to many data acquisition needs.

P & D Analytics Incorporated

2. Business Terminology

Good communication depends on each party understanding the terms and references used. The following definitions guide the reader on some important terms and also how this book utilizes these terms.

Root cause analysis terminology:

Fault tree: A very useful graphic tool that identifies the top failure and then expands using both "OR" and "AND" logic gates until key components are identified as primary factors resulting in the top event or failure.

Fishbone: A graphic tool deconstructing a failure into four or five key components categories. Some examples are People, Process, Material, and Management. This tool helps the investigator successfully explore these common categories. The fishbone diagram does not contain any logic to its branches, whereas the fault tree identifies the logic related to the gates.

Proximate cause: The proximate cause is normally described as the key component in a system that failed. *It is not the root cause.* For example a component failing in a system is the proximate cause of the failure.

Root cause: The key primary cause of an issue. For example, the failed process resulted in a component being installed in a device. The failed process was the root cause of the failure.

Business Analysis terminology:

Constraint: A restriction or choke point in a process. This is the step that controls the process speed and output.

Extrapolation analysis: A mathematical analysis using existing X and Y data (Y is the dependent variable) to predict a value Y1, given an X1 which is outside of the range of existing values. The analyst needs to be cautious regarding extrapolation, because the relationship may be very different outside the known range.

Feedback control: This method adjusts the input into a system based on the output. In other words, it can scale back on the input in order to prevent an overshoot.

Interpolation analysis: A mathematical analysis, using existing X and Y data, to predict an intermediate value Y1 given an X1.

Mathematical emulation model of electrical or mechanical system: This describes the modeling of how a system will work or respond to assumed conditions. A model could either be performance based models or reliability/risk emulation of systems.

Pivot table: A pivot table is an EXCEL® spreadsheet function that enables users to analyze the relationship between multiple categories of information in a large data set and thereby estimate the relational significance between categories. Users commonly refer to the function as "slicing and dicing" the data set to understand the relationship between specific parameters.

Regression analysis: A mathematical method using several data points to predict where another point falls on the line that represents the best fit to the set of data points.

Requirement: A specification that must be met to satisfy the objectives of a process or product.

Reliability: The probability that a system or process will meet its objective or perform properly.

Risk: A predictive tool based on the likelihood and consequence resulting from a set of events or a specific scenario. It is a predictive measure of the probability of various scenarios.

These terms and tools guide decisions and methods for managing everyday business. Although process owners may not always realize it, they are using these tools informally to make decisions each day.

Measurement terminology:

Parameter measurement is a complex subject. The following text is not an in-depth description of key parameter measurement concepts. However, a basic understanding is needed before the measurement practitioner proceeds. These concepts are important because they can impact the value of the measurements obtained.

Accuracy: Parameter indicating whether a measurement device will read the correct value. It also describes the performance of a measurement device. *Accuracy* and *precision* are sometimes incorrectly used to describe the same aspect, but they are different as noted here, and both are important.

Arithmetic mean or average: Mathematically calculated by dividing the sum of a set of values by the number of those values.

Direct measurement: This means the actual parameter of interest is measured directly, for example, the number of times an event occurs or the distance traveled.

Inferential measurement: This term means the parameter of interest is being calculated or in some other fashion inferred from a different measurement. Some physical parameters are very difficult to measure, but can be calculated by measuring another parameter when a relationship is known. A *resistive thermal device* or RTD has resistance changes based on temperature, so the resistance is measured and the temperature is inferred.

Linear or nonlinear relationships: A linear relationship indicates that the equation that matches the data is in the form of $Y = MX + B$, where M is the slope and B is the Y axis intercept. The Y value also varies directly with the X value. A *nonlinear relationship* has an equation that matches the data in form of $Y = MX^C + B$, where M is a factor in the curvature of the line. C is also a factor which significantly effects the curvature of the line. B is the Y intercept. EXCEL® has a built-in function to quickly calculate and display both linear or nonlinear relationships.

Noise: In the case of measurement, noise represents a variable reading at the low end of the scale being measured. If the readings are in this area, they may be overwhelmed by the noise and make it difficult to determine what is actually happening.

Percentage of full scale: Some instrumentation accuracy is provided as a percentage of full scale. Normally this is in the upper end of the measurement range.

Precision: Measure indicating how repeatable a system is when multiple measurements are taken. It is normally used to describe the performance of a measurement device.

Standard deviation: A measure of the variance about the mean.

Uncertainty: Represents a measure of aspects that either can't be measured or may impact the results of a process but are difficult to control.

This book focuses on capturing data to assist in a root cause analysis, but once you have data it can be utilized in other areas to improve process performance. *P & D Analytics* is developing a subsequent book, which will describe how to utilize data to minimize constraints, develop root cause strategies, and find pivotal requirements so that the business can grow into a learning organization.

Electrical Terminology:

The following terms are key for understanding the Arduino UNO and the sensors it utilizes.

Baud rate: The rate at which data is transferred across the serial port. 9600 Baud is 9600 bits per second.

IDE: Integrated Development Environment is the tool for generating code for the Arduino and uploading it to the device.

Microcontroller: Device providing control and data transfer capabilities. It has limited programming capability. Commands are uploaded that tell the microcontroller what to do. The Arduino is a microcontroller.

Pull-up Resistor: Resistor providing the pullup resistance to increase the voltage difference from the open position of a switch to the closed position. This ensures the Arduino "sees" this difference.

Resistor: A resistor is a device that provides resistance in a circuit. One good use for a resistor is to compare a known resistance to a variable resistance. This can tell the microcontroller what has changed on a sensor output.

Serial port: A serial port is a connection to a computer, normally using a USB connector. The Arduino connects using the USB serial port.

Transistor: A transistor is a device that has an input, output, and signal line. It acts like a gate. As the signal changes it allows current to flow from input to output. This is very helpful as a sensor because changes in the signal line allow current to flow through the transistor.

Wheatstone bridge: A circuit that contains four resistors connected end to end. Input voltage is applied at two connections located across from each other. If one of the resistors changes resistance slightly, a differential voltage will be seen at the other nodes. It is a small output voltage, so normally an amplifier is needed to increase the voltage to a level that is easily measured.

P & D Analytics Incorporated

3. Basics of Root Cause Analysis

Root Cause Analysis

Root cause analysis refers to a multi-step process which consists of the following:

- Identify potential causes related to a significant issue.
- Determine potential corrective actions.
- Identify and implement corrective actions.
- Monitor the process to determine if the corrective action(s) eliminated or minimized the significant issue.

Many methods and tools exist to implement this process. Some characteristics that lead to a successful root cause analysis are: an open-minded team, management support, good facilitation skills, key data, creative ideas to solve the problem, and following the proper sequence of steps. The steps below outline a general root cause analysis process and its implementation in more detail then the four steps noted above:

1. *Problem occurs* and is significant enough to warrant a formal root cause analysis team to be established.

2. *Set up a root cause team*

- Team leader
- Facilitator/root cause analyst
- Key technical experts
- Management involvement at the start and at key milestones

3. *Hold initial kickoff meeting.* This important meeting educates everyone on the process and methods that will be utilized along with developing a definition of the problem. Using a brainstorming technique, develop the first draft list of causes. Finally, ensure the analysis scope is defined.

4. *Investigate the causes* and expand the cause list to ensure that all the realistic potential causes are included. The use of a fishbone diagram, fault tree, or some other device is an excellent method to capture, document, and communicate these causes.

5. *Hold an evidence meeting.* After key causes have been identified, gather the key evidence data that exists. Determine what other evidence is needed.

6. *Gather additional evidence.* The material in this book can assist this phase. If more data are needed, an Arduino system may be the low-cost and timely solution to capture additional data and evidence.

7. *Determine the most likely causes* as a team, using all of the evidence (and not speculation). A good way to ensure the evidence makes sense is to write out the description using the "If____ then____" format. This ensures the existence of an authentic cause-effect relationship between the evidence and effect.

8. *Hold a corrective action meeting,* in which management and the team lead engage to ensure that the corrective actions are feasible and to emphasize the need to implement solutions. Each corrective action should be assigned to an individual with a defined closure date.

9. *Monitor the process* to ensure that the corrective actions are implemented and resolve the problem and have not created new ones.

The process described above is idealized. It is best not to jump to conclusions. However, human nature tends to want to solve problems based on minimal evidence and/or "gut" feel. It is wise to stay in the evidence gathering stage as long as needed before jumping to the corrective action phase. Conversely, by the time the team has been assembled, there may have been several attempts to "solve" the problem. If so, this may provide additional data for the root cause analysis team to evaluate.

The authors suggest *using a schedule* to ensure that the evidence gathering phase is given enough time to produce real evidence. Setting up a few basic Arduino systems ahead of a problem can help capture the data quickly and efficiently. This book can help plan what types of sensors might be needed, so that potentially useful sensors can be kept in stock, ready to go.

Example Problem

In order to provide guidance on potential uses of the Arduino, this book will use an example problem to develop measurement methods of various physical parameters. These measurements will be used to gather evidence on an example product that is exhibiting a large number of defect reports.

Problem Statement:

A company manufactures an enclosure (Figures 1 and 2) which holds an electronic system, and it has exhibited moisture leaks into the enclosure while in a hot, humid environment. The electronics case is sealed with an O-ring in a captured volume (See Figure 2). The product defects found include scratches and paint that bridge across the O-ring sealing surface and could be leakage paths for moisture to enter the enclosure.

For this example, let us assume the defect rate is relatively high and has garnered management attention. A root cause analysis team has been set up and given some time to investigate the problem. Additional information regarding the enclosure and seal arrangement is in the figures below.

Figure 1: Similar Seal Area in Example Problem

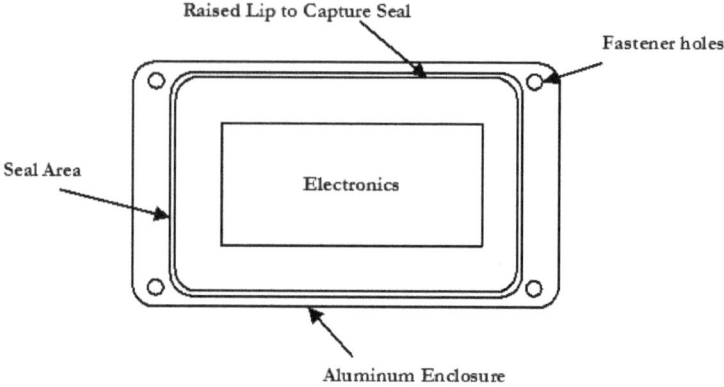

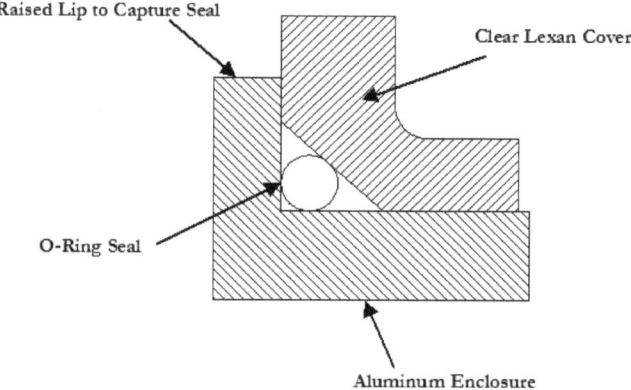

Figure 2: Example Enclosure Seal Top View and Cross Section

After brainstorming, the root cause analysis team developed the following diagram referred to as a "fishbone" because of its fishbone appearance. The fishbone diagram is a tool commonly used in root cause studies. It assists the team by defining several generic problem categories that may include the cause of the problem. In this example the team determined that Process, People, Design, and Environment are the general causal categories. The next step in root cause analysis for the team is to gather evidence that either validates a cause or eliminates it. Figure 3 below shows the fishbone diagram that the team developed.

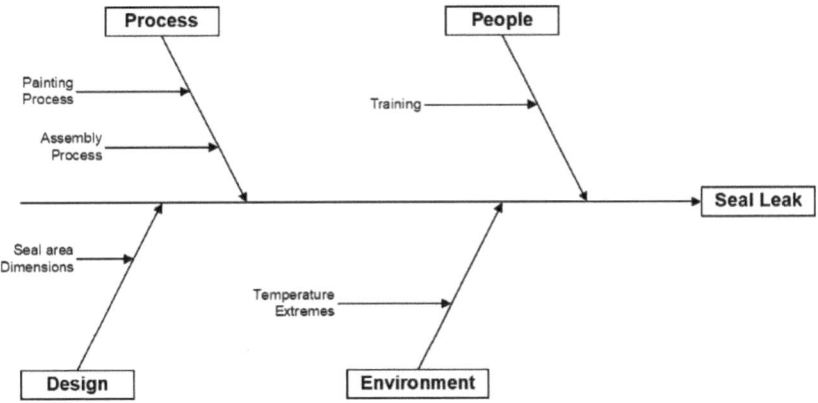

Figure 3: Example Fishbone Diagram from Root Cause Analysis

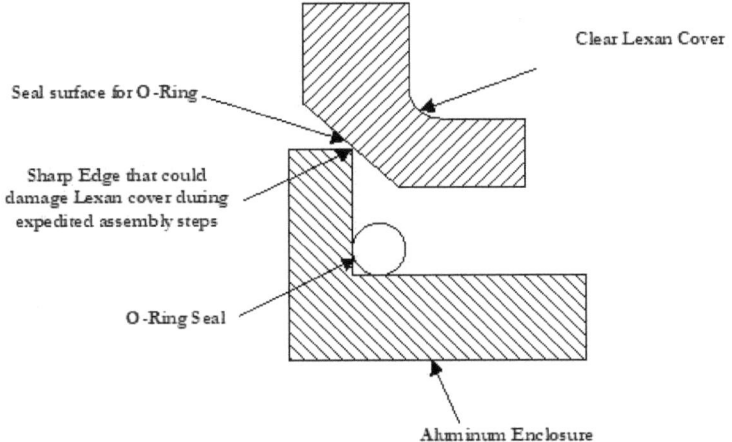

Figure 4: Diagram Of Expedited Assembly Damage

Figure 4 highlights the concern regarding the expedited assembly process, as the cover is installed after the subassembly to move it to final assembly, and when the cover is installed in final assembly.

As noted above, the team used the fishbone diagram to determine potential causal categories. These categories raised questions associated with the process steps, shown in Figure 5. This illustrates how to probe one step deeper at each process step and how the cause could come from that step.

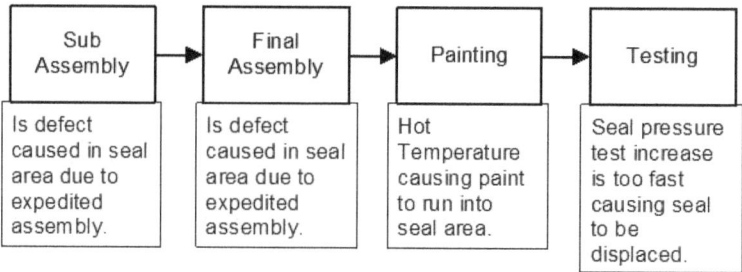

Figure 5: Process Step diagram with defined questions

Resource [5] Bonner highlights how temperature can affect the curing of UV coatings. This relates to the concern about temperature causing the paint to run into the seal area.

The next subsection elaborates on the temperature extremes in the Environment branch of the fishbone diagram.

Temperature Extreme due to Environment Analysis

From the fishbone diagram above the team identified that the extreme temperature in the field, along with general design considerations associated with the seal, might be the cause of the leak. This is due to the thermal growth of the Lexan cover compared to the growth of the aluminum enclosure. The Lexan has approximately twice the thermal growth rate of aluminum. The calculation below in Table 1 and shown in Figure 6 is based on the team's assumption that there is a temperature change of approximately 60 degrees F between night and the heat of the day. If this is the case, then it might result in the Lexan cover growing so much that the compression on the O-ring would decrease enough that it would no longer seal. One other design aspect considered is that the Lexan cover seal area is a cast-in feature, so there is an additional tolerance related to it. With this additional cast tolerance and the temperature growth rate it may be hard to maintain the O-ring compression.

If the investigator wants to understand these aspects in greater detail, please see the sources in Appendix Section H: Material Properties, Casting Tolerance, and O-Ring Compression Information.

Table 1: Calculation of Thermal Growth and Potential Seal Gap

Calculation of difference between Lexan and Aluminum Thermal Growth	
Lexan Coefficient of thermal expansion in/ (in °F)	3.75E-05
Aluminum Coefficeint of thermal expansion in/ (in °F)	1.31E-05
Difference between Coefficients of therm expansion in/ (in °F)	2.44E-05
Length inch	5
Difference Temperature between set up and high temp (130 - 70) = 60 °F	60
dl inches = α * L * (T1 - T2) =	0.0073

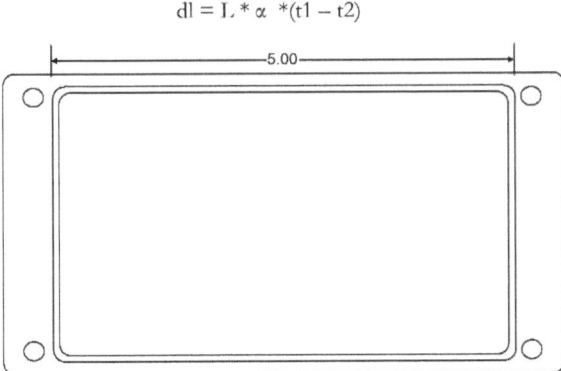

$$dl = L * \alpha * (t1 - t2)$$

Figure 6: Seal Dimension Used in Thermal Growth Calculation

To answer the question regarding the 60°F day to night temperature change, a system to measure this needs to be developed.

Again, this is an example to show how to use the Arduino to gather key data needed to resolve the issue. The root cause analysis team gathers data on each potential causal question, either validating the cause or eliminating it. The causes, evidence, and recommended mitigations should be documented in a report to ensure the same issues can be identified quickly and resolved in the future.

The next chapters examine how to use the Arduino to gather and assesses the data to answer the Figure 5 questions, along with the questions related to design and environment.

P & D Analytics Incorporated

4. Measurement, Arduino UNO, Data Capture/Analysis, and Safety

Measurement

For the novice in measurement, this section provides guidance on how to get started and answers many basic questions. If the reader is a test or measurement engineer or technician this may provide refresher information or new insight into the Arduino.

What is measurement and why is it important to measure process parameters in the business world?

Measurement is the use of technology and physical methods to gather information that is difficult to get via our normal senses. This information may be critical because it can provide insight into how failures occur, or how a business operation is performing or changing over time. It may also provide insight into making choices at key decision points. In other words, it provides valuable data to aid our decisions. Without these data, decisions can be based primarily on guesswork and may result in subpar performance or even induce severe consequences.

What happens when we rely on guesswork, and we are successful?

Do we become overconfident in our ability to predict the future? In many cases that is exactly what can happen. Then when our guesswork decision process breaks down, bad things can happen, hopefully without catastrophic results.

The guidelines in this book demonstrate relatively easy technical solutions to gain insight into basic parameters for understanding potential causes affecting a product or a process. The projects in this book are relatively straightforward methods to gain insight into some basic physical and operational functions.

What are some examples of key process measures that could result in product problems?

- *Time* it takes to complete a process step.
- *Variability* in final product due to time differences between the end of one step in a process and the start of second step.

- *Temperature* changes at a critical time in the process.
- *Light intensity* affecting temperature.
- *Pressure change rate* above specifications.

This book shows how to develop systems to measure these and other parameters using a unique microcontroller tool that connects directly to a computer. It is called an **Arduino UNO**, and it can send data directly to the computer over a USB port.

Arduino UNO

The Arduino is a relatively new microcontroller development board that is ready to program and acts as an interface between a personal computer and various devices and sensors. It measures and controls many different items well, and is perfect for developing small measurement systems.

Brief History and General Description

The original Arduino board was first developed in 2004 as a tool to help students learn programming. It is an open-source tool (which means using the code and system configurations is free as long as credit is provided) and has developed a large base of helpful websites and user groups. It represents a breakthrough as an easy-to-use, relatively inexpensive, programmable interface The development tool to program it can be downloaded free from the Arduino website.

The Arduino [3], Adafruit [1 and 2], Sparkfun [9], and other websites are great places to start learning the potential of the Arduino tool. Several introductory books also help the researcher get started using this device. *Getting Started With Arduino* by Banzai Resource [4] is a very good beginner's book on Arduino. Other sources of information for the Arduino novice are the maker fairs and user group activities.

Even though the Arduino was created as a training and hobbyist tool, because of its options, versatility, ease of programming, and large number of online resources, it is a viable option for serious applications. The authors have run an Arduino UNO system frequently for more than two years. It has operated flawlessly. Even with this operational experience the Arduino UNO should not be considered a permanent measurement system.

Arduinos come in several versions and sizes. The projects in this book utilized the **Arduino UNO Rev3**. The IDE (Integrated Development Environment) programming tool used is **version 1.8.1**. The coding language used by the Arduino is very similar to C+.

A larger version of the Arduino board exists called the *Mega*. It has 54 digital input and output ports and 16 analog input pins. The same IDE tool is used to program this device. The Mega board might be the perfect choice if you plan to measure many inputs. The Arduino UNO has not changed much in several years, so previous work, schematics, and information are still usable and up to date. Even though many different clones and configurations exist, the authors recommend starting with an authentic Arduino. This ensures that it will work well and be compatible with the sensors and code.

Note: The authors found that when setting up an Arduino it is advantageous to mount it to a small wooden block. This allows the addition of a proto-board or terminal strip or securing a secondary device relative to the Arduino for wiring purposes. The parts list for each project does not contain the wooden block, terminal strip or proto-board and any additional fittings or tubing connectors that may be seen in the figures. It shows one example of how to configure the system. The designer may want to configure it in a different way to fit their unique situation.

Description of Arduino UNO Ports and Interfaces

Figures 7 and 8 below depict the main Arduino UNO ports.

Figure 7: Arduino UNO

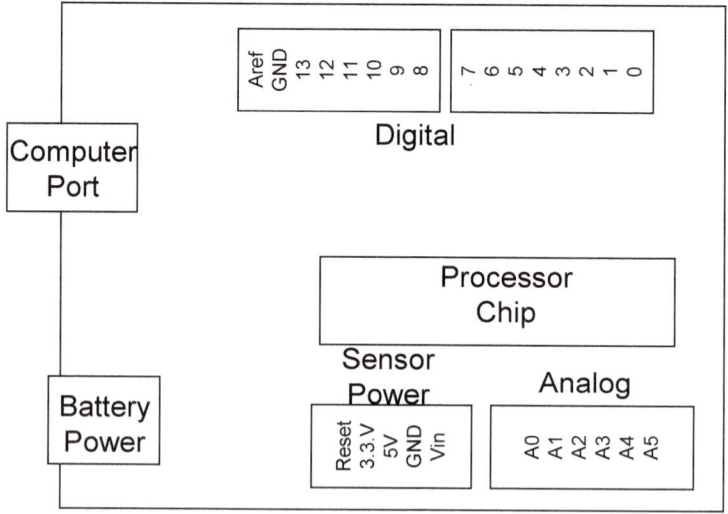

Figure 8: Arduino UNO Ports

Five primary port groupings are used to connect to the Arduino.

Computer port: This is the primary port that connects directly to the computer. It is used to upload programs (referred to as "sketches") to the Arduino and carry the data to the computer.

Battery power: This port allows an Arduino to be unplugged from a computer and use a 9 to 12 Volt DC adaptor or 9 Volt battery power to perform Arduino operations.

Sensor power port: These port connections provide 3.3 Volt and 5 Volt DC power for sensors or other devices, like motors. A limited amount of current can be provided.

Analog device ports: These header connections enable analog inputs.

Digital device ports: These header ports enable digital inputs and outputs.

Sketch/IDE: The IDE is the programming tool that runs on a computer for writing code (i.e., the *sketch* in the Arduino world). Once developed in the IDE, the code is uploaded to the Arduino. As noted above, the IDE tool must be downloaded from the Arduino website.

Shields/breakout boards: These are add-on boards that are either inserted into the standard Arduino board ports or connected via wires.

Libraries: These are non-volatile packages of software that are meant to be called upon many times throughout different programs. Non-volatile in this context means not changing much, if at all. The Arduino allows the use of the library the same way as before, despite having several other components changed over time. However, over time these libraries can be updated and changed. You may need to check the folders located in both of these paths to ensure that you have the correct libraries:
- *document > arduino > libraries*
- *%arduino installation path% > libraries locations*

Data Capture and Logging

Several unique but inexpensive methods can be used to capture and store data which may be critical in the root cause/corrective action process. The easiest method shown in this book simply uses the Arduino UNO to send data over the USB serial data port directly to a computer. It can be copied using *Control C* after it is highlighted and then pasted into a spreadsheet or other document.

Another method stores data remotely on an SD card. See Appendix for information on how to build the Adafruit Data logging Shield. This board allows the user to set up the Arduino so it saves the data on an SD card, which can be removed from the Arduino Shield to download the data to the computer. Once downloaded to the computer, the data can be analyzed using tools like an EXCEL® spreadsheet.

The following sections describe the process of using the Arduino UNO to gather data.

Using the Serial USB Port to log data from the Arduino

Once the Arduino is running, it is easy to access the data from its USB connection to the computer. Simply click on the button on the Arduino Development tool as seen in Figure 9 below:

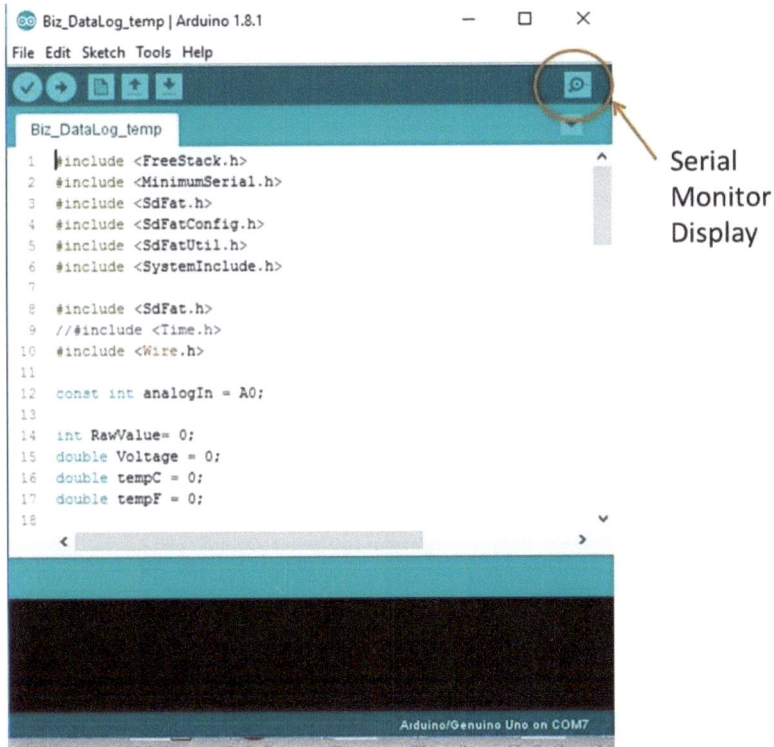

Figure 9: Arduino IDE Serial Port Button

The following figures exemplify the type of data and information sent over the serial port to the computer using Arduino as the bridge. The investigator can tailor the information inside the Arduino program so that the important data can be selected, formatted, and sent over the serial line to the computer. Figure 10 shows an example of data coming across the serial port.

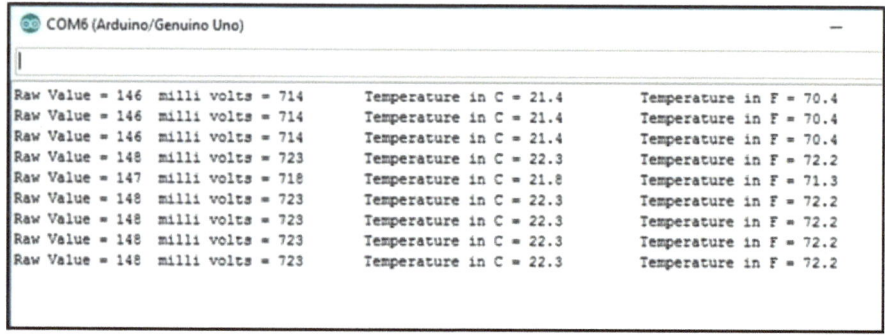

Figure 10: Example Serial Data from Arduino

This information can be copied, pasted into a spreadsheet, and displayed as a graph (See Figure 11 below).

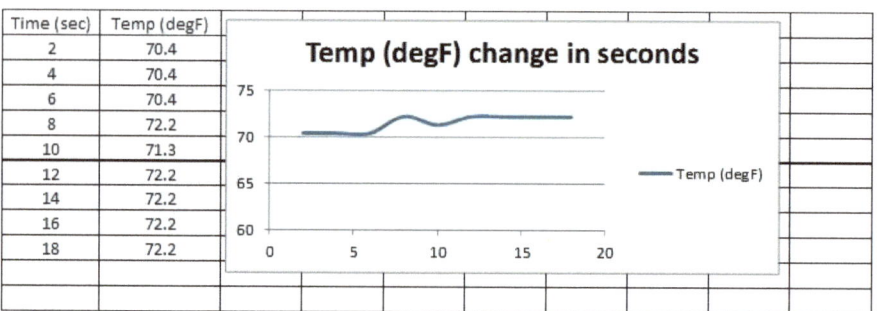

Figure 11: Graph of Example Serial Data

Data Analysis Concepts

Constraint Analysis

There are many types of data analysis tools. One that is very powerful are the functions built into EXCEL® and other spreadsheets. The following graphic uses data measured or captured regarding an operational process. It can establish where constraints control the total output of the process. These graphics are powerful tools that aid data visualization and direct changes needed to minimize the constraint impact. Using the Arduino tools described in this book, the process owner can rapidly upload the data to a computer where it can be inserted into a spreadsheet to quickly update graphics like the one in the Figure 12.

Subassembly A step 1		Subassembly A step 2			
Time: 15 minutes		Time: 10 minutes			
A Step 1 Parts Needed	4	A Step 2 Parts Needed	3		
A Step 1 Parts Available	20	A Step 2 Parts Available	12	Capacity A	
Capability Subassembly A	5	Capability Subassembly A	4	4	

				One A and One B needed for C	
Subassembly B step 1		Subassembly B step 2		Constraints prior to Assembly C	
Time: 20 minutes		Time: 5 minutes		Time (minutes)	30
B Step 1 Parts Needed	8	B Step 2 Parts Needed	3	Capability for day	3
B Step 1 Parts Available	30	B Step 2 Parts Available	15	Capacity B	
Capability Subassembly B	3.75	Capability Subassembly B	5	3	

Figure 12: Example Constraint Diagram

The diagram above shows a snapshot of an assembly process that takes component A and B to assemble the final product C. The graphic highlights that from a time perspective, Step 1 of Subassembly Part B takes 20 minutes, and Step 2 of Subassembly A takes 10 minutes. This results in the potential maximum constraint of 30 minutes to complete both Sub-Assemblies. Most likely there would be some overlap and the time might be reduced, but there is still the potential for dead time resulting in 30 minutes.

The next feature that this diagram shows is that the limit on total assemblies is 3, because there are only enough parts to finish 3 Sub-Assemblies through step 1. This type of analysis can be very useful for finding and mitigating the impacts of constraints.

Mathematical Relationships

Another powerful function in EXCEL® is the ability to graph data and visualize the relationship between independent variables and a dependent variable. This relationship can be expressed in an equation called a regression, which produces an R^2 value. Figure 13 shows an example. A *perfect relationship* between independent and dependent variables would be expressed as a sloping line (regression line) with *all data points on the line (no deviations at all)*. The equation value (referred to as R^2) in this perfect-fit case would yield a value of 1.0. Such relationships in engineering and science seldom occur. More commonly, the data appear on a graph as a cloud of points scattered above and below the regression trend line, and the R^2 value that the equation yields is something less than 1. A positive R^2 value indicates an upward trend, whereas a negative R^2 value indicates a downward trend. Generally, the closer the R^2 value is to 1.0, the better the data fit the derived equation (See figure 13 below). The test engineer or investigator should note however that an R^2 value near 1.0 only points to a *potential cause-effect relationship* between variables and not necessarily an actual meaningful relationship. It does point to something that the investigator may or may not decide to investigate further.

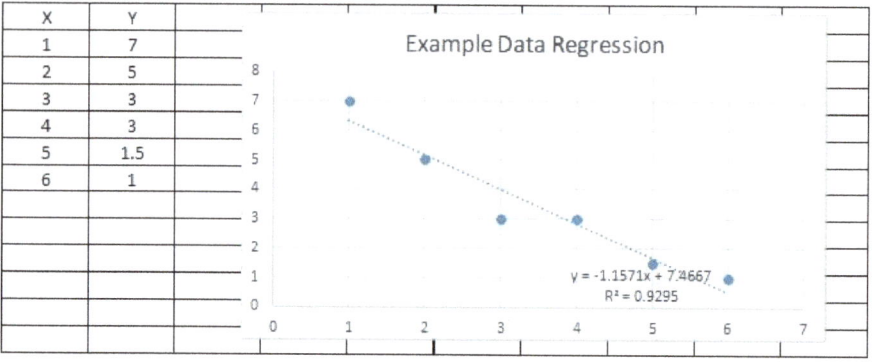

Figure 13: Linear Regression graph example

To display the equation and the R^2 after setting up the graph, click on the line, select *format trend line*, then make sure *Display equation and Display R squared value* are checked. If the investigator believes a cause-effect relationship exists, plug a value (or values) for X (i.e., the independent variable) into the equation to calculate the value of Y (i.e., the dependent variable).

There are other techniques and tools, but always a good start is to graph the data to visualize the relationships and see if any patterns exists.

All systems have inherent errors related to measuring any parameter. A method to assess this is explained in the Appendix F Uncertainty Analysis. These sensors and the Arduino should not be considered a standard, but rather a working measurement device. Appendix F outlines the standard method for calculating measurement uncertainty. If the investigator is interested, the following source provides extensive information related to measurement and calibration.

https://www.nist.gov/services-resources/standards-and-measurements

Safety Information

Please note the following safety aspects:
- Do not utilize the Arduino or any of its sensors in a hazardous environment. They are not designed for hazardous applications, and they could ignite a flammable gas.

- Also, when working with 110 Volts or higher, there is the potential for shock. The Arduino is designed for low voltage, but always be careful around any wires that may be powered and exposed.

- When soldering, always wear eye protection and try not to breathe the fumes. Remember, a recently soldered connection can be very hot.

Other Information

Arduino is a great way to automate data entry into computers. Don't let setting up and programming an Arduino be discouraging. The Arduino is relatively easy to program. Additionally, the large user community and lots of examples online are helpful. A great starting place is www.arduino.cc. Finally, there are many fairly low-cost sensors available that easily interface with the Arduino and can be used to build powerful data collection systems.

Some other options and vendors such as *Lab View, Fluke, and Omega Engineering* provide advanced tools. These may provide excellent solutions, but might be expensive and difficult when determining which system best fits your objectives. The Arduino might be used as a test to guide the analyst or test engineer's decision on which advanced solution is best for their situation.

Finally, projects in this book are probably not the exact solution for the test engineer's situation. More likely, these ideas may inspire the development of other tools and techniques based on systems presented here. The authors hope these tools provide the information needed to solve your tough measurement problems.

These next chapters highlight several projects that utilize Arduino systems to answer the questions posed in the example root cause analysis.

5. Measuring Counts and Timing of Operations

This chapter starts the process of tracking down the cause of the seal leak from the root cause analysis example in section 2.

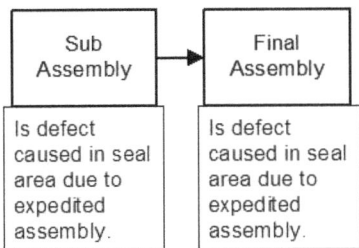

Figure 14: Example Problem Assembly Step Questions

One of the key questions from the root cause analysis of the example problem relates to the seal area being damaged due to expedited assembly operations. Data may not exist on how much time it takes to complete the subassembly and the final assembly. An estimate of the time may be calculated based on the total number of units completed in a day, thereby determining if the assembly processes are being expedited, resulting in damage to the seal area on the Lexan cover.

However, if that rough estimate is not adequate for determining if that step is being rushed, then the three projects below can provide more precise values of the time it takes to complete steps.

The first project is pretty basic and can provide more information related to the number of counts or times an operation occurs. It uses a simple switch to capture an event and sends this count over the serial line to the IDE. This information can be captured by the investigator and copied and pasted into a spreadsheet to be analyzed.

The second project refines the first system by using timing pulses to capture the state of the switch every ten seconds. The code has just a few added steps from the previous project, but provides a way to estimate the actual time the switch is closed.

The third project is more complex and utilizes a separate board called a Real Time Clock. This captures the time from the computer and utilizes that to time tag when the data is sent. It will provide a more precise measure of the time between events.

The investigator will need to determine how precise an estimate is needed to discover whether the operations are being expedited, and how to set up the switch to capture the data. In these projects the assumption is that the switch will be placed in such a way that when the enclosure is set down on the work bench it activates the switch. See the diagram below.

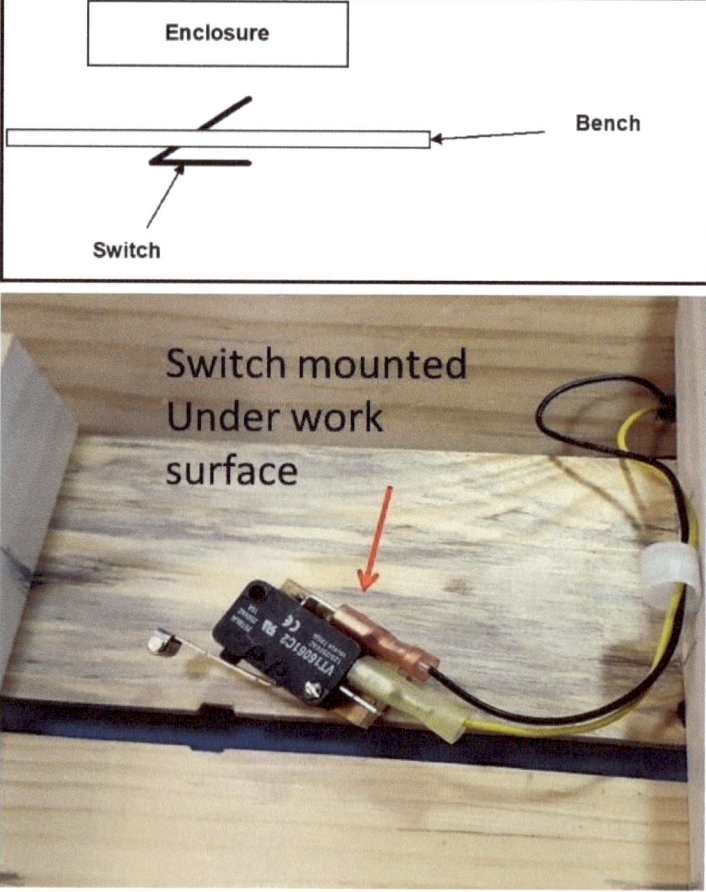

Figure 15: Switch Under Work Surface

The investigator may choose to utilize other methods for switch closure.

Project 1: Arduino UNO SN001

This project provides a very simple solution that counts operations by switch closure and sends this data to a computer. Once in the computer the data can be copied and pasted into a spreadsheet. This allows analysis of event timing for the assembly operations in the root cause example.

Description/Goal: The first system is relatively straightforward and uses an Arduino to capture counts with a switch connected to the digital port (See Figures 15 and 16). The total number of switch closures can be divided by the total time, giving an average time per step. This can provide data to determine if the assembly operations are being expedited, causing damage to the seal area. The goal is to demonstrate how to build a system that will track counts using switch closures and to learn about the Arduino.

Challenges: The first challenge is getting a simple Arduino system running with this great introductory project. One other challenge is setting up a switch that will activate when needed.

Hardware needed:
- Arduino UNO
- Computer
- Serial connection cable between computer and Arduino
- Switch: Single Pole Single Throw (SPST) momentary contact switch. The version the authors used is a lever activated switch.
- 10K ohm resistor (This is the external pull-up resistor which ensures the Arduino can detect the switch closure).

Figure 16: Arduino and Switch

The components are connected per the schematic below (Figure 17).

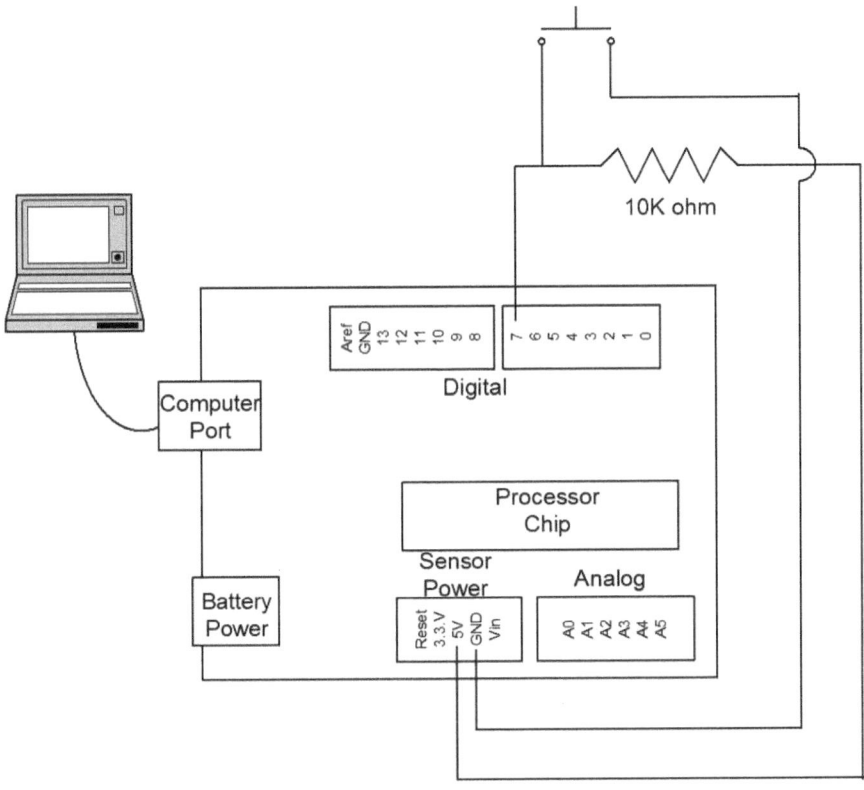

Figure 17: SN001 Counting Schematic with External Pull-up Resistor

This code is uploaded to the Arduino via the IDE.

```
//SN001_Counter_only_apr29_2017
//Code developed by P & D Analytics Inc.

int counter = 0;
//Program uses Switch to set high state and increments count
//holds high until it changes to low state.
//Sends count increment over serial line

bool switchOn = false;

void setup() {
  Serial.begin(9600);
  Serial.println("This is SN001 Program by P&D Analytics");

}

void loop() {

//The next step increments the counter every time switch changes

  if(digitalRead(7) == HIGH and switchOn == false){

    Serial.print("Switch has been turned on for ");
    Serial.print(counter);
    Serial.print(" times.");
    Serial.println(" ");
    counter = counter + 1;
    switchOn = true;

//The delay step ensures switch does not bounce
    delay(500);
  }

//The next step determines the state of the switch
  if(digitalRead(7) == LOW and switchOn == true){

    switchOn = false;
  }
}
```

Getting the system up and running.

The first step is to label the Arduino UNO as SN001 (or whatever numbering configuration the test engineer determines). This helps tie the code to the hardware. Upload the code via the IDE.

The program consists of three primary elements or functions.

1. The first important element of code block checks the input for the initial position of the switch, and establishes the baud rate to communicate with the computer over the serial port.

2. One line prints the code name to the serial port to ensure help identify the correct code.

3. The next block of code is the first "if" in the loop. and it monitors the switch position. When the switch is closed the Arduino sends that information and the count number to the computer via the serial port.

The following switches are some other options but might require a bit of adjustment to ensure they activate when the enclosure is placed on top of it.

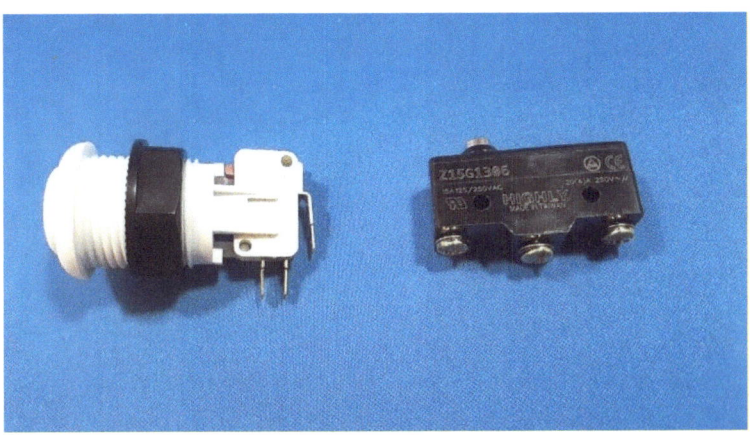

Figure 18: Other Optional Switches

Once you have the counts and timing between events, how do you use it?

Observing trends or patterns that may occur over the course of a day. This can be used to track expedited assembly trends. Expedited assembly trends that coincide with defective products may provide evidence that this could be a root cause. See graph below:

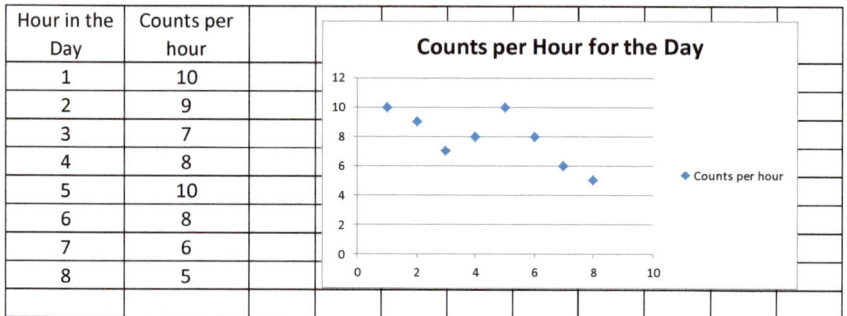

Hour in the Day	Counts per hour
1	10
2	9
3	7
4	8
5	10
6	8
7	6
8	5

Figure 19: Counts per Hour Trends

Another method of utilizing data is interpolation, a technique in which a midpoint is calculated based on existing data. The most common method is linear interpolation between two points.

Beyond this root cause analysis there is another potential use for count data, which is to extrapolate near-term trends based on existing data. Using the data for the first four weeks and the trend line equation function in EXCEL® the investigator can assess if there are other trends occurring that may be serious enough to warrant mitigation. See graph below.

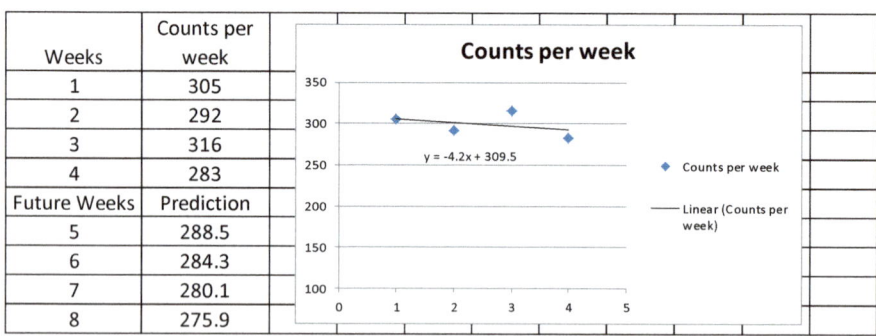

Weeks	Counts per week
1	305
2	292
3	316
4	283
Future Weeks	Prediction
5	288.5
6	284.3
7	280.1
8	275.9

Figure 20: Counts per Week

As always more data provide better insight, but it is important to *ensure a trend is real* and not just part of normal variation.

Getting back to the example root cause analysis, this information can be used to find trends with specific assembly steps that may result in damage to the seal area. If there are trends within an assembly step, it might indicate which step needs to be investigated in more depth, or point to some sort of protection that may be needed over the seal during that operation or do not install cover until final assembly and do it carefully. Tracking time between events can provide valuable information on any process. It can also help with "debottlenecking" a process or removing constraints.

Project 1A: Arduino UNO SN001A

This project builds on the previous project by slightly different approach that will provide a relatively accurate estimate of the time a switch is closed. This data is sent over the serial port and can be copied and pasted into a spreadsheet.

Description/Goal: This system uses the same hardware as SN001 where the only change is to code to capture an estimate for how much time the switch is closed and open. The investigator will learn about modifying existing code.

Challenges: This has the same challenge as the previous project, which is setting up the Arduino and the switch that will activate properly to capture the timing of events.

The primary difference between this code and the first one is that a signal is sent over the serial line every ten seconds that shows the state of the switch. Using this data, a rough estimate of the time the switch is closed can be developed by adding up the number of ten second increments. If the ten second increment is too short, increase the delay to whatever is needed. This will increase the coarseness of the resolution but still provide useful data. This is a modified version of the DigitalReadSerial code that is included as an example when you download the Arduino IDE. The file is located in the example folders located where the Arduino code is stored on your computer. The investigator may find other interesting example programs in this folder that might be useful.

Code for capturing estimated time that switch is closed on SN001A.

```
/*
SN001A Counts and Time estimate
Modified DigitalReadSerial which is in the public domain.
Reads a digital input on pin 7, prints the result to the serial monitor
at 10 second intervals.
*/

// digital pin 7 has a pushbutton attached to it. Give it a name:
int pushButton = 7;

// the setup routine runs once when you press reset:
void setup() {
  // initialize serial communication at 9600 bits per second:
  Serial.begin(9600);

  Serial.println("This is SN001A Program by P&D Analytics.");

  // make the pushbutton's pin an input:
  pinMode(pushButton, INPUT);
}

// the loop routine runs over and over again forever:
void loop() {
  // read the input pin:
  int buttonState = digitalRead(pushButton);
  // print out the state of the button:
  Serial.print("State of Switch is ");

  //Using Exclusive OR to get reverse state indication.
  int displayState = (buttonState ^= 1);
  Serial.print(displayState);

  delay(10000);      // delay in between reads
  Serial.println(" for 10 seconds ");
}
```

Getting the system up and running.

The first step is to label the Arduino UNO as SN001A. This helps tie the code to the hardware if the question ever comes up. Upload the code via the IDE.

1. The program consists of two primary elements or functions. The first element of code names the switch and pin for inputs, sets up the routine, and sets up the baud rate to communicate with the computer over the serial port.

2. The next block of code captures the state of the switch every ten seconds and sends that information over the serial port. The code reverses the output using an Exclusive OR in order to get a 0 for switch when it is open and 1 for closed.

The following image shows the output of the serial port.

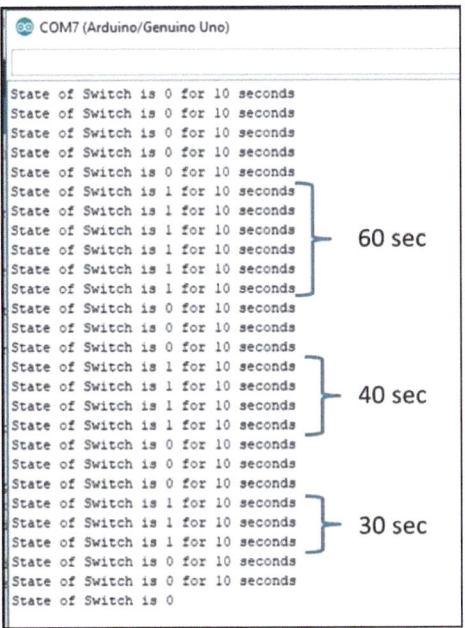

Figure 21: Serial Output from SN001A

From the data shown above there is a trend toward speeding up the event as the time the switch is closed has changed from 60 to 40 to 30 seconds.

Project 1B: Arduino UNO SN001B

This project takes the system one step further and utilizes a Real Time Clock to capture more accurate timing when a switch is closed.
For more information on the Real Time Clock breakout board the website below explains it in detail. It also provides the starting point for the code utilized to capture data.

http://bildr.org/2011/03/ds1307-arduino/

Description/Goal: The Arduino in this project gathers time data from the Real Time Clock module then sends that over the serial port only when the switch is activated or closed. It checks every second. The investigator can use that to determine how long it takes to complete an event. The investigator will learn how to use breakout boards and other elements of the Arduino microcontroller.

Challenges: Setting up the Real Time Clock and the switch to activate when needed to capture real data can present some challenges.

Hardware needed:
- Arduino UNO
- Computer
- Serial connection wire to connect to the Arduino from the computer
- Switch: any type of Single Pole Single Throw (SPST) momentary contact switch. The version the authors used is a lever activated switch.
- 10K pull-up resistor
- Real Time Clock Module (Sparkfun Part # BOB-12708)
  Wires were soldered into the board to connect to the Arduino.

The components are connected per the picture and schematic below (Figure 22 and 23).

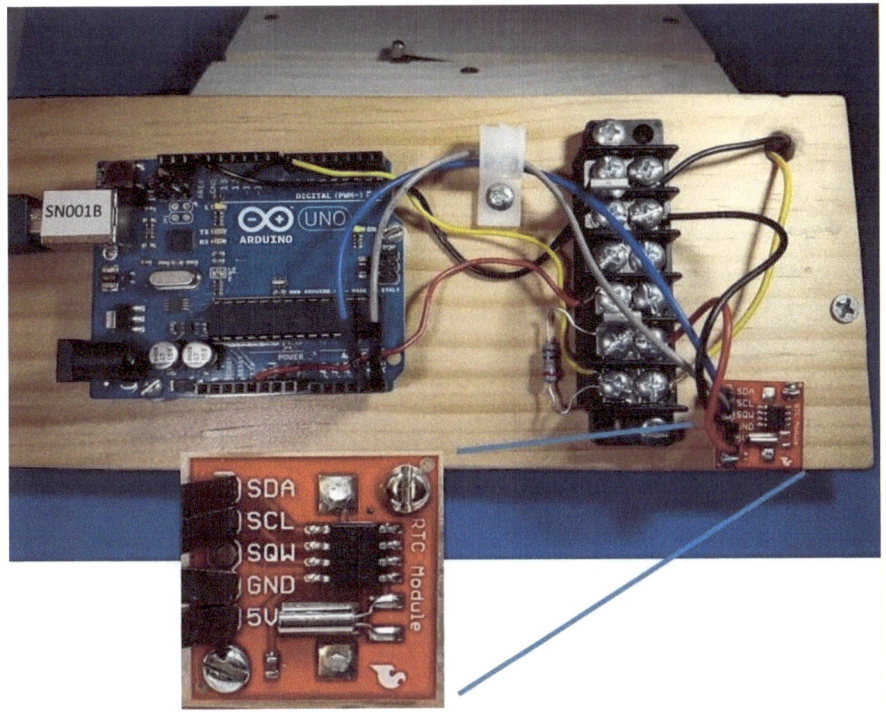

Figure 22: SN001B Counts with Real Time Clock

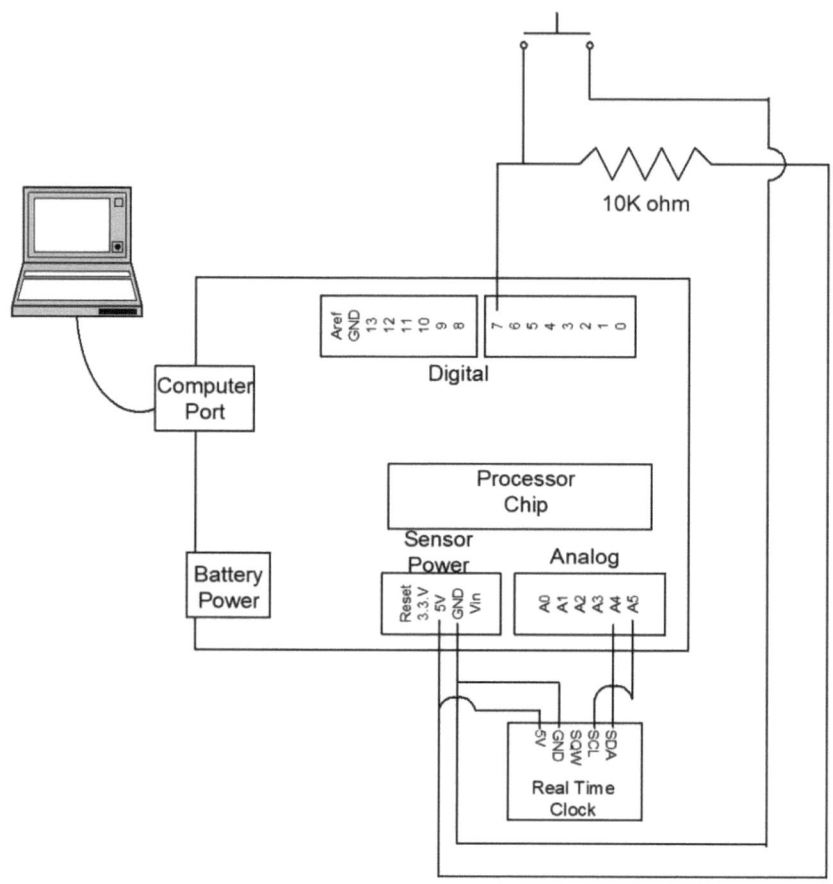

Figure 23: SN001B Counts with Real Time Clock Schematic

There are two libraries that need to be added to this project: Wire.h and SparkfunDS1307RTC.h.

Code for the Counts using the Real Time Clock SN001B project.

```
//SN001B_RTC_July_16_2017
#include <Wire.h>
#include <SparkFunDS1307RTC.h>
#define DS1307_ADDRESS 0x68
#define SQW_INPUT_PIN 2
#define SQW_OUTPUT_PIN 13
```

```cpp
int second = 0;
int minute = 0;
int hour = 0;
int weekDay = 0;
int monthDay = 0;
int year = 0;
int month = 0;

void setup() {
  Wire.begin();
  Serial.begin(9600);

  Serial.println("This Program is SN001B_7_16_17 by P&D Analytics");

  pinMode(SQW_INPUT_PIN, INPUT_PULLUP);
  pinMode(SQW_OUTPUT_PIN, OUTPUT);
  digitalWrite(SQW_OUTPUT_PIN, digitalRead(SQW_INPUT_PIN));

  rtc.begin();

  rtc.writeSQW(SQW_SQUARE_1);

  //rtc.setTime(s, m, h, day, date, month, year);
  //day = day of the week; date = day of the month
  //Need to reset line below to current date!!!!
  rtc.setTime(30, 35, 7, 1, 16, 7, 17);
}

void loop() {
  if(digitalRead(7)==LOW)
    printDate();

  delay(1000);
}

byte bcdToDec(byte val){
  return ((val/16*10) + (val%16));
}

void updateTime(){
  rtc.update();
  Wire.beginTransmission(DS1307_ADDRESS);
```

```
  byte zero = 0x00;
  Wire.write(zero);
  Wire.endTransmission();

  Wire.requestFrom(DS1307_ADDRESS, 7);

  second = rtc.second();
  minute = rtc.minute();
  hour = rtc.hour();
  weekDay = rtc.dayStr();
  monthDay = rtc.date();
  month = rtc.month();
  year = rtc.year();
}

void printDate() {
  char *temp = malloc(2);

  updateTime();

  //Print the date in this format 3/1/17 23:59:59

  sprintf(temp, "%02d", month);
  Serial.print(temp);
  Serial.print("/");
  sprintf(temp, "%02d", monthDay);
  Serial.print(temp);
  Serial.print("/");
  Serial.print(year);
  Serial.print(" ");
  sprintf(temp, "%02d", hour);
  Serial.print(temp);
  Serial.print(":");
  sprintf(temp, "%02d", minute);
  Serial.print(temp);
  Serial.print(":");
  sprintf(temp, "%02d", second);
  Serial.println(temp);
}
```

Getting the system up and running.

The first step is to label the Arduino UNO as SN001B. This helps tie the code to the hardware if the question ever comes up. Upload the code via the IDE.

The program consists of four primary elements or functions.

1. The first lines of code calls up the libraries. These files need to be located in the library subfolder in the Arduino folder on the computer.

2. The next steps starts the clock. *It is important to change the date to the current date here as it sets the clock to the correct time.* Otherwise you may get a strange date in your output.

3. The next step sets the switch up for acquiring counts and then acquires the date information when a switch is closed.

4. The final section of code sends the date information over the serial bus.

How to use the data.

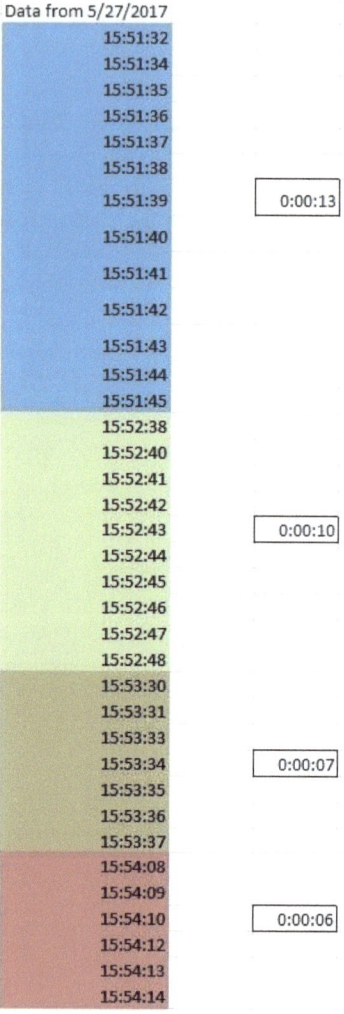

Figure 24: Data from Real Time Clock SN001B Pasted into Spreadsheet

When the switch was activated, the Arduino started sending data over the serial line that had the time of day in a month : day : year : hours : minutes : seconds format. The Arduino keeps sending the data until the switch is opened. The data from the serial output was then copied and pasted into a spreadsheet. In the figure above each unique event was color coded, so it is easy to see when it started and when it finished. Then it is simply a matter of subtracting the starting time of each event from the finish time.

The next step is to graph the data and observe the trend. From the collected data, there is less time being spent on this step. Therefore, this could be the cause of the damaged seal area.

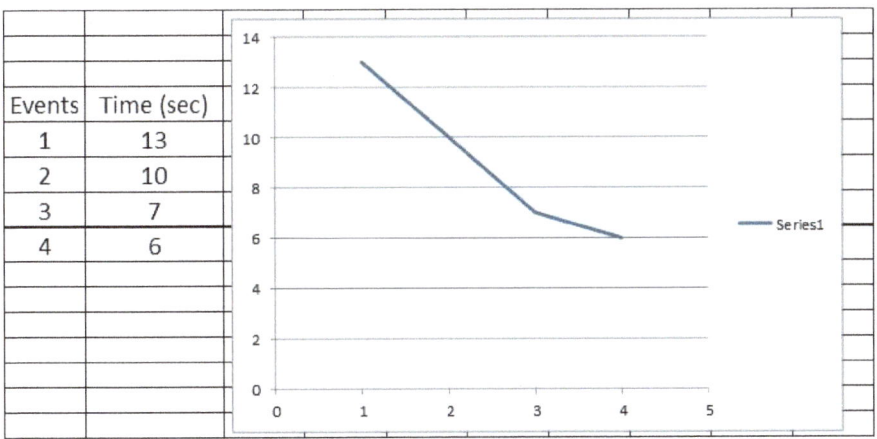

Events	Time (sec)
1	13
2	10
3	7
4	6

Figure 25: Trend of Event Timing Using Data from SN001B

These three projects provide methods to capture the number of counts and with that determine if there are issues with expedited assembly operations that need to be investigated and mitigations developed by the root cause analysis team.

P & D Analytics Incorporated

6. Measuring Temperature

This chapter highlights methods to measure air temperature and its effect on drying paint. This was one question the root cause analysis team developed during the brainstorming session, because temperature can affect many materials and processes. The team believed this could be a cause due to the principle that heat normally speeds up chemical reactions and reduces viscosity. As a result high temperature can cause paint to flow more easily into other areas and then dry quickly.

Could temperature have caused the wet paint to flow into the seal area?

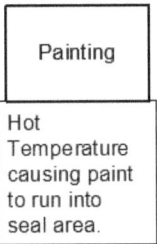

Figure 26: Painting Step and Temperature Drying Question

There are four projects in this chapter, and they all provide methods to measure temperature. Three use one type of sensor, and the last one uses a different sensor. The first project is basic, and each subsequent project either adds capabilities or other interfaces.

The first project uses a very common, inexpensive temperature sensor, the TMP 36. The investigator can find many examples online that show how to use this device. The code used in this project sends the data over the serial port to the computer.

The second project uses the same sensor, but adds an Arduino Shield which will save the temperature data on an SD card. This allows the investigator to set up the Arduino system using a power supply and remove the computer if needed. The Arduino will capture data and write it to the SD card until powered off.

The third project uses the first system, but sends some of the data to a Liquid Crystal Display (LCD) so that it is visible at that local sensor location. The data are also sent over the serial port to the computer.

The final project uses a different temperature sensor called a "one-wire" device. It can be purchased packaged in a water tight casing. Several of these devices can be connected to the "one-wire" signal line, and the information is sent over the serial port to the computer.

Project 2: Arduino UNO SN002

This project develops a basic system, but can provide a very useful method to capture temperature data and send it over the serial port.

Description/Goal: The system uses the Arduino and its analog signal inputs by connecting a temperature sensor to it. This can measure temperature over time, and the Arduino will send the data over the serial port to the computer. Another goal is to learn how to solder and build sensor packages.

How does the temperature sensor work?

The temperature sensor is an *TMP 36 transistor*. A transistor acts as a gate with one input opening or closing the gate, thereby regulating the amount of electrical current flowing through it. For the TMP 36 the change in temperature controls either an increase or decrease in the electrical flow.

Challenges: Ensuring the temperature sensors are calibrated and protected from the environment (See Appendix for assembly of temperature sensor into a robust device and how to calibrate it).

Hardware needed:
- Arduino Uno
- Computer
- Temperature sensor TMP 36 (This sensor can be powered with either the 3.3 Volt of 5 Volt ports on the Arduino.)

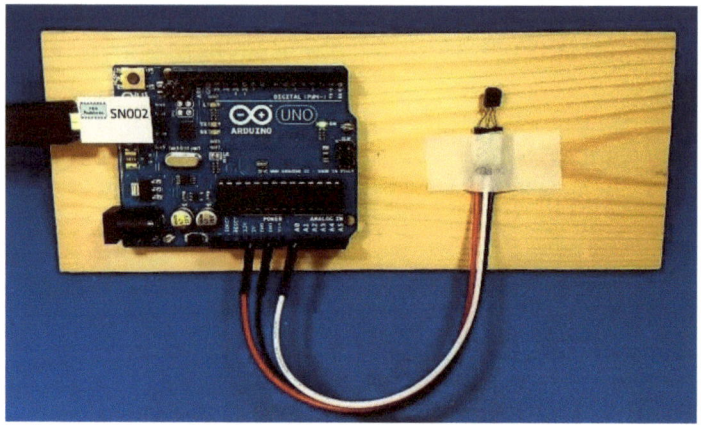

Figure 27: SN002 Simple Temperature

The figure below shows the sensor's flat side. In the schematic (Figures 29) the sensor's flat side is coming out of the page. It is important to get the orientation correct.

Figure 28: Flat Side of TMP36 Sensor

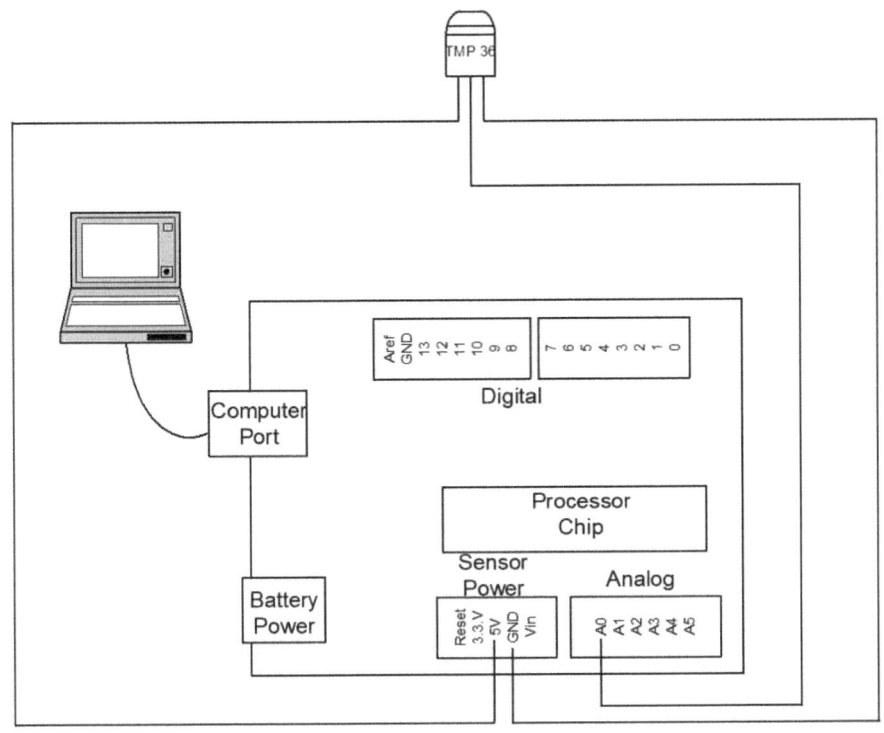

Figure 29: Schematic for SN002 Temperature Measurement

Many examples of Arduino code for using the TMP 36 temperature sensor are available online. A very good website with a lot of information related to this sensor is:

http://henrysbench.capnfatz.com/henrys-bench/arduino-temperature-measurements/simple-tmp-36-arduino-thermometer/

SN002 code is a slightly modified version from the website above.

```
/*
SN002 The simplest TMP 36 Thermometer
This code is from henrysbench.capnfatz.com
*/
const int analogIn = A0;

int RawValue= 0;
double Voltage = 0;
double tempC = 0;
double tempF = 0;

void setup(){
  Serial.begin(9600);

  Serial.println("This is SN002 Program by P&D Analytics");
}

void loop(){

  RawValue = analogRead(analogIn);
  Voltage = (RawValue / 1023.0) * 5000; // 5000 to get millivots.
  tempC = (Voltage-500) * 0.1; // 500 is the offset
  tempF = (tempC * 1.8) + 32; // convert to F
  Serial.print("Raw Value = " ); // shows pre-scaled value
  Serial.print(RawValue);
  Serial.print("\t milli volts = "); // shows the voltage measured
  Serial.print(Voltage,0); //
  Serial.print("\t Temperature in C = ");
  Serial.print(tempC,1);
  Serial.print("\t Temperature in F = ");
  Serial.println(tempF,1);
  delay(500);
}
```

Getting the system up and running.

The first step is to label the Arduino UNO SN002. The next is to upload the code via the IDE to the Arduino.

The code is made of two primary elements

1. The first part of the code sets the input values.
2. The next part of the code captures the data and sends it over the serial port to the computer every half second.

The figure below shows how the data looks as it comes across the serial port and also how it looks when you highlight it prior to using "Control C" to copy it .

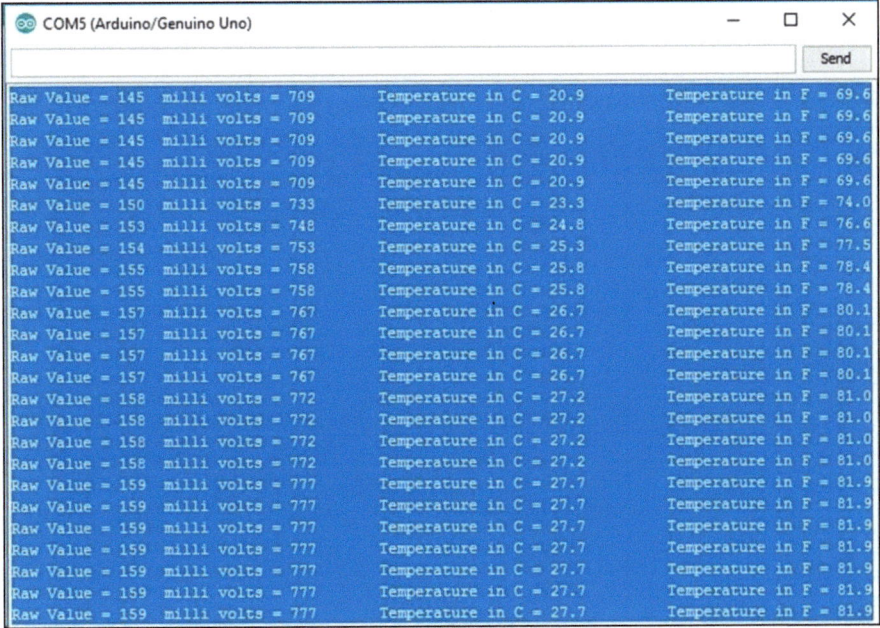

Figure 30: Example SN002 Data from Serial Port

How to use the data.

The following graph shows how to analyze the data to monitor the temperature trends and evaluate if it could cause the paint to run into the seal area. This information shows that the temperature is increasing. This could be enough to cause the paint to migrate into the seal area. See reference [5] for considerations associated with temperature and paint.

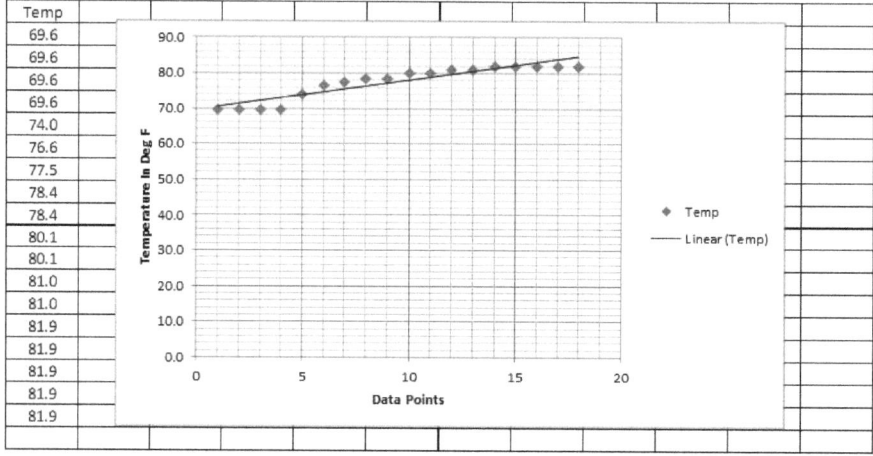

Figure 31: Analysis of SN002 Temperature Data

Project 2A: Arduino UNO SN002A
Using the TMP36 Temperature sensor and SD card Data Logger shield

This section expands the temperature and paint in the seal question one step further. The Arduino captures data over a lengthy period of time and allows the investigator not to tie up a computer during that period. It combines a temperature sensor with an Adafruit Data Logging Arduino Shield (See Figures 32 and 33). The data logger has an SD card where the data are saved. With a separate power supply this device can be set remotely away from your computer. It counts the time and, at prescribed intervals, takes a temperature reading and writes that data to the SD card. This might be beneficial in measuring the change in temperature of a certain location when an external event occurs. Adding another sensor provides that ability to measure stratification aspects of other locations in the room.

Description/Goal: The investigator will learn to utilize the powerful Data Logger Adafruit Arduino Shield.

Challenges: This Adafruit Data Logger Shield requires some assembly and is much more complex code, but it will help to make the Arduino stand alone and operate without a computer connected to it.

Hardware needed:

- Arduino Uno
- Computer
- Temperature sensor TMP 36
- Adafruit Data Logger Shield Part # 1141 (The authors utilized different connector headers than are supplied with the kit. The Adafruit site shows how to add these headers, also see Appendix E.)
- 9 volt power supply once programming completed

Figure 32: Data Logging System on Remote Power Supply

How to build the system

The following schematic (Figure 33) shows how to connect the temperature sensor to the combined Arduino and Data Logger.

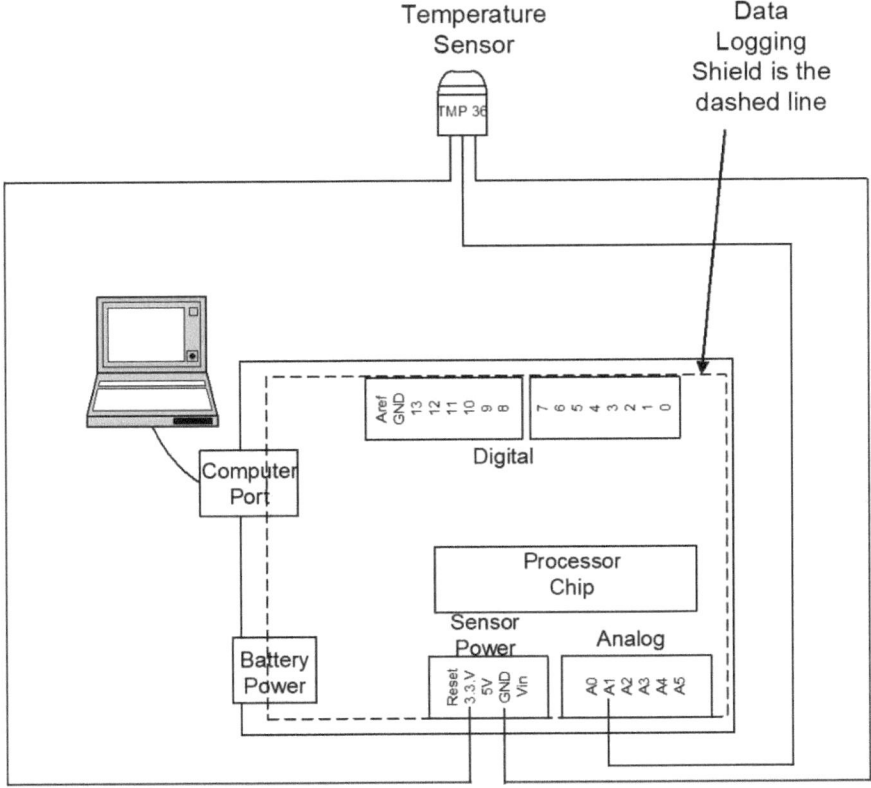

Figure 33: Schematic of SN002A Data Logging System

Add libraries to get the code to operate properly

This system requires a number of libraries to be added to the library folder on the computer. The libraries are normally located in the documents folder. These libraries are like subroutines that are called up when needed and reduce the amount of code that the investigator needs to program. For this code the libraries needed are: SD.h, Wire.h, and RTClib.h.

The SN002A Code is modified from some the code listed on the AdaFruit website. The website is very comprehensive and can provide more guidance if the investigator would like to gain greater understanding of how to use this it by reviewing the instructions here:

https://learn.adafruit.com/adafruit-data-logger-shield/using-the-real-time-clock-3

This is the SN002A code and is based on the elements from the website above.

```
//SN002A_SD_card_July_19
#include "SD.h"
#include <Wire.h>
#include "RTClib.h"

// A simple data logger for the Arduino analog pins
#define LOG_INTERVAL  1000 // mills between entries
#define ECHO_TO_SERIAL   1 // echo data to serial port
#define WAIT_TO_START    0 // Wait for serial input in setup()

// the digital pins that connect to the LEDs
#define redLEDpin 3
#define greenLEDpin 4

// The analog pins that connect to the sensors
#define tempPin 1          // analog 1
RTC_PCF8523 RTC; // define the Real Time Clock object

char daysOfTheWeek[7][12] = {"Sunday", "Monday", "Tuesday", "Wednesday", "Thursday", "Friday", "Saturday"};

// for the data logging shield, we use digital pin 10 for the SD cs line
const int chipSelect = 10;

// the logging file
File logfile;

char filename[] = "LOGGER00.CSV";

void error(char *str)
{
```

```
  Serial.print("error: ");
  Serial.println(str);

  // red LED indicates error
  digitalWrite(redLEDpin, HIGH);

  while(1);
}

void setup(void)
{
  Serial.begin(9600);
  Serial.println();

  Serial.println("This Program is SN002A by P&D Analytics");

#if WAIT_TO_START
  Serial.println("Type any character to start");
  while (!Serial.available());
#endif //WAIT_TO_START

  RTC.begin();

  if(! RTC.initialized()) {
    //Serial.println("RTC is running!");
    RTC.adjust(DateTime(F(__DATE__),F(__TIME__)));
  } else {
    //Serial.println("RTC is NOT running!");
  }

  // initialize the SD card
  Serial.print("Initializing SD card...");
  // make sure that the default chip select pin is set to
  // output, even if you don't use it:
  pinMode(10, OUTPUT);

  // see if the card is present and can be initialized:
  if (!SD.begin(chipSelect)) {
    Serial.println("Card failed, or not present");
    // don't do anything more:
    return;
  }
  Serial.println("card initialized.");
```

```
  for (uint8_t i = 0; i < 100; i++) {
    filename[6] = i/10 + '0';
    filename[7] = i%10 + '0';
    if (! SD.exists(filename)) {
      // only open a new file if it doesn't exist
      logfile = SD.open(filename, FILE_WRITE);
      break; // leave the loop!
    }
  }

  if (! logfile) {
    error("couldnt create file");
  }

  Serial.print("Logging to: ");
  Serial.println(filename);

  Wire.begin();
  if (!RTC.begin()) {
    logfile.println("RTC failed");
#if ECHO_TO_SERIAL
    Serial.println("RTC failed");
#endif //ECHO_TO_SERIAL
  }


  logfile.println("millis,time,light,temp");
#if ECHO_TO_SERIAL
  Serial.println("millis,time,light,temp");
#endif // attempt to write out the header to the file

  pinMode(redLEDpin, OUTPUT);
  pinMode(greenLEDpin, OUTPUT);

  // If you want to set the aref to something other than 5v
  //analogReference(EXTERNAL);

  logfile.close();
}

void loop(void)
{
```

```
  // delay for the amount of time we want between readings
  delay((LOG_INTERVAL -1) - (millis() % LOG_INTERVAL));
  
  digitalWrite(greenLEDpin, HIGH);
  
  logfile = SD.open(filename, FILE_WRITE);
  // log milliseconds since starting
  uint32_t m = millis();
  logfile.print(m);         // milliseconds since start
  logfile.print(", ");
#if ECHO_TO_SERIAL
  Serial.print(m);          // milliseconds since start
  Serial.print(", ");
#endif

  // fetch the time
  DateTime now = RTC.now();
  
  // log time
  
  logfile.print(" since midnight 1/1/1970 = ");
  logfile.print(now.unixtime()); // seconds since 2000
  logfile.print(", ");
  logfile.print(now.year(), DEC);
  logfile.print("/");
  logfile.print(now.month(), DEC);
  logfile.print("/");
  logfile.print(now.day(), DEC);
  logfile.print(" ");
  logfile.print(daysOfTheWeek[now.dayOfTheWeek()]);
  logfile.print(" ");
  logfile.print(now.hour(), DEC);
  logfile.print(":");
  logfile.print(now.minute(), DEC);
  logfile.print(":");
  logfile.print(now.second(), DEC);
  
#if ECHO_TO_SERIAL
  Serial.print(" since midnight 1/1/1970 = ");
  Serial.print(now.unixtime()); // seconds since 2000
  Serial.print(", ");
  Serial.print(now.year(), DEC);
```

```cpp
    Serial.print("/");
    Serial.print(now.month(), DEC);
    Serial.print("/");
    Serial.print(now.day(), DEC);
    Serial.print(" ");
    Serial.print(daysOfTheWeek[now.dayOfTheWeek()]);
    Serial.print(" ");
    Serial.print(now.hour(), DEC);
    Serial.print(":");
    Serial.print(now.minute(), DEC);
    Serial.print(":");
    Serial.print(now.second(), DEC);
#endif //ECHO_TO_SERIAL

    delay(10);
    int tempReading = analogRead(tempPin);

    // converting that reading to voltage, for 3.3v arduino use 3.3
    float voltage = (tempReading * 5.0) / 1024.0;
    float temperatureC = (voltage - 0.5) * 100.0 ;
    float temperatureF = (temperatureC * 9.0 / 5.0) + 32.0;

    logfile.print(", ");
    logfile.println(temperatureF);

#if ECHO_TO_SERIAL
    Serial.print(", Temperature Measure in F: ");
    Serial.println(temperatureF);
#endif //ECHO_TO_SERIAL

    digitalWrite(greenLEDpin, LOW);

    logfile.close();
}
```

Getting it up and running

Label the Arduino UNO SN002A. Load the code to the Arduino using the IDE.

This complex code has five main elements.

1. The first element sets up all of the parameters and loads the libraries.

2. The second element sets up the SD card and which pin for the SD card line.

3. The next element sets up the baud rate for the serial port, and also starts the process of watching for data and ensuring the SD card is working properly.

4. The next element sends data over the serial line. If a computer is not planned to be connected, these elements could be commented out by putting two // in front of the line of code not used. However, it does not matter and the program will function fine if they are left in.

5. The final element writes the data to the SD card.

One difference between this and the previous projects is that the data is stored on a file on the SD card. It is text data and can be copied pasted into a spreadsheet. In EXCEL© select "paste special" and then "text" to automatically put each data item in its own cell. Some of the cells will contain both numbers and text. The text in the cell can be deleted so that you only have numbers.

Project 2B: Arduino UNO SN002B
This project adds an LCD display to the temperature measurement system, which will display the temperature information captured by the Arduino. It still sends the data over the serial port to the computer for analysis later.

Description/Goal: This uses the code from the SN002 system but adds additional code to send data to the LCD. This can provide insight at or near the location of the temperature sensor and still send the data over the serial port to store on the computer, too. The investigator will learn to use an LCD in conjunction with the Arduino.

Challenges: The authors have had some challenges with the LCD device but the code below seemed to work very well.

Hardware needed:

- Arduino Uno
- Computer
- Temperature sensor TMP 36
- Sparkfun LCD Part # LCD-09395

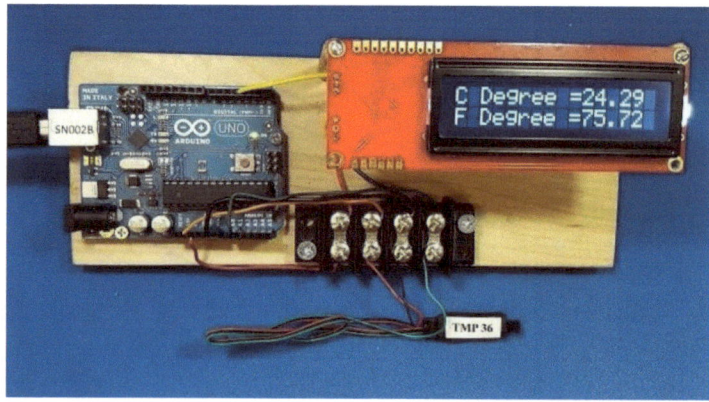

Figure 34: LCD and TMP 36 Temperature Sensor

How to build system

Connect the Arduino, sensor, and LCD per the schematic below.

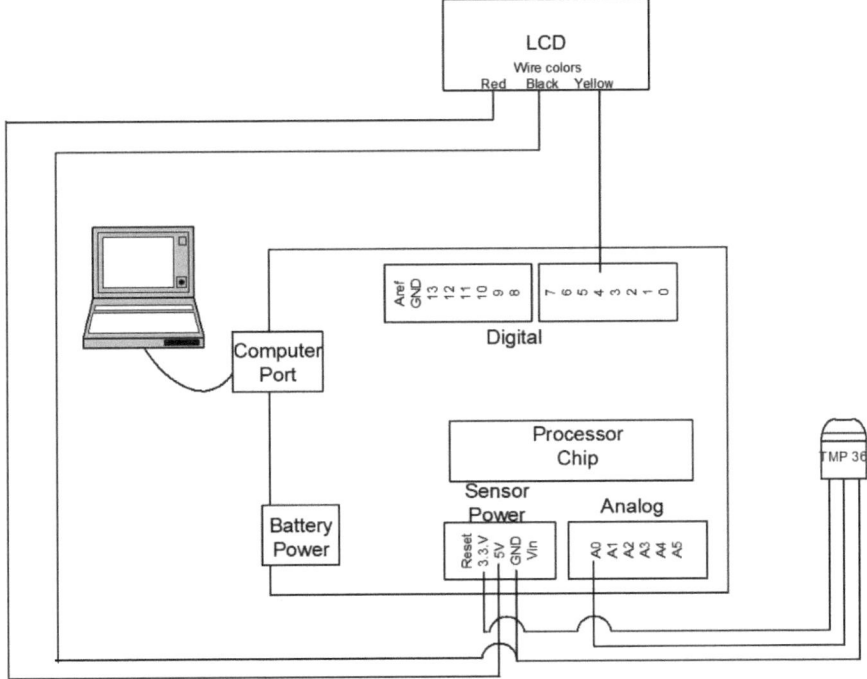

Figure 35: SN002B LCD and Temperature Sensor

Upload the code to the Arduino via the IDA.

The SN002B code is modified version from the Sparkfun code for the LCD. The SoftwareSerial.h library will need to be loaded.

```
//SN002B_Temp_LCD_5_16_2017
//Original code from:
// https://www.sparkfun.com/tutorials/246

#include <SoftwareSerial.h>

SoftwareSerial softwareLCD(3,4);
```

```
const int analogIn = A0;

int RawValue = 0;
double Voltage = 0;
double tempF = 0;
double tempC = 0;

void setup() {
  Serial.begin(9600);

  Serial.println("This is SN002B Program by P&D Analytics");

  softwareLCD.begin(9600);
}

void loop() {
   RawValue = analogRead(analogIn);
   Voltage = (RawValue/1023.0) * 5000; //5000 for millivolts.
   tempC = (Voltage-500)*0.1; //500 is offset
   tempF = (tempC*1.8) +32; //Converting to F degree

   Serial.println(tempC,1);

   softwareLCD.write(254);
   softwareLCD.write(128);

   softwareLCD.write("                "); // clear display
   softwareLCD.write("                ");

   softwareLCD.write(254);
   softwareLCD.write(128);

   softwareLCD.write("C Degree =");
   softwareLCD.print(tempC);
   softwareLCD.write(254);
   softwareLCD.write(192);

   softwareLCD.write("F Degree =");
   softwareLCD.print(tempF);

   delay(500);
}
```

Getting it up and running

Label the Arduino UNO SN002B. Upload the code to the Arduino using the IDE.

The code consists of four elements.

1. The first element sets up the input port and initiates the Software Serial library.

2. The second element attaches the LCD to the its ports.

3. The next code elements increases the resolution of the variables using the double parameter.

4. The final section reads all the data and sends is to both the serial port and the LCD.

How to use the data.

The only difference from the first project is the use of an LCD display located near the source of the data. This might provide insight regarding high temperatures and indicate the need to reduce the temperature. The effect of the temperature reduction will be tracked in the data coming across the serial port.

Project 2C: Arduino UNO SN002C

This project utilizes a different sensor type to measure temperature It has one main advantages over the TMP36 sensor, but is a little more complicated to use.

The sensor utilized is *DS18B20 sensor* (Figure 36) which can be purchased in a version that is encased in a waterproof package. This might be necessary if the temperature sensor will be located in a wet environment.

Description/Goal: The type of sensors utilized in this project are designated one wire. This might seem confusing since they come in configurations of three and sometimes two wires. The reason for this designation is that many sensors can be connected to the same signal bus, i.e., the same single wire.

Challenges: The authors have sometimes had problems with one wire devices, but the code below seems to work very well. You will need to have both the one wire and the Dallas Temperature library installed.

Hardware needed:

- Arduino Uno
- Computer
- 4.7 K ohms resistor
- DS18B20 one wire temperature sensor

Figure 36: DS18B20 One Wire Temperature Sensor

How to build the system

Connect the Arduino, sensor, and resistor per the schematic below.

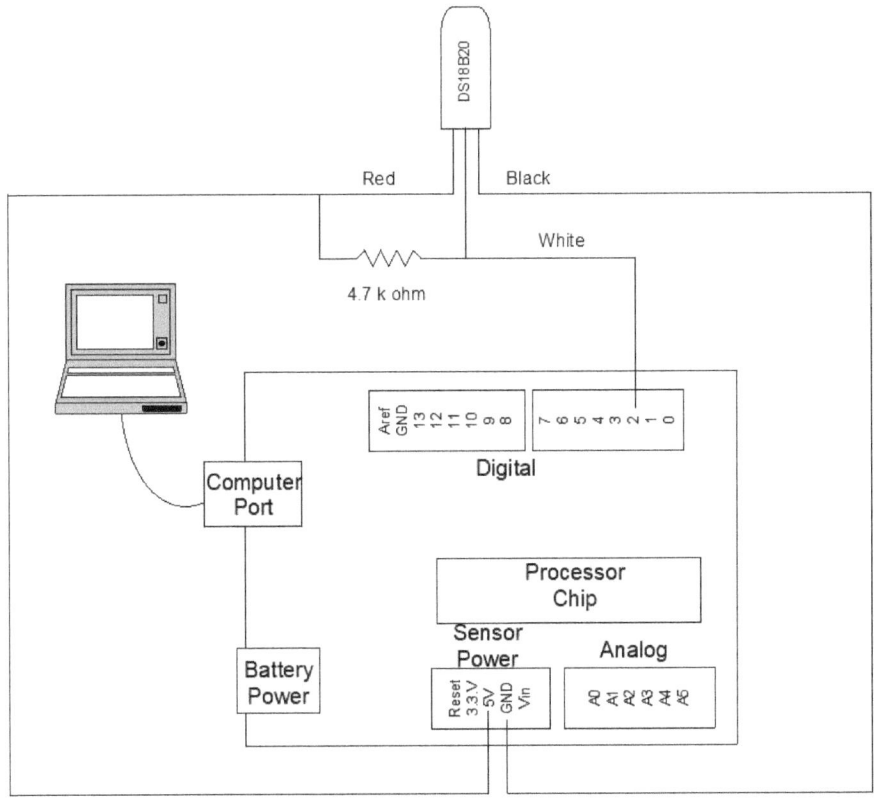

Figure 37: SN002C Schematic DS18B20 Sensor

The following code can be used to read the sensors and is located on the website listed. This website provided a lot of good information related to this temperature sensor. The code below comes directly from the website with only modifications to some of the comment lines. The DallasTemperature.h and OneWire.h libraries will need to be loaded.

https://create.arduino.cc/projecthub/TheGadgetBoy/ds18b20-digital-temperature-sensor-and-arduino-9cc806

Code for SN002C project, load it with the IDE. This project needs the OneWire.h and DallaTemperature.h libraries.

```
//SN002C_1wire_5_19_2017
// Include the libraries
#include <OneWire.h>
#include <DallasTemperature.h>
#define ONE_WIRE_BUS 2

//Configure Wire Bus
OneWire oneWire(ONE_WIRE_BUS);

//Pass our oneWire reference to Dallas Temperature.
DallasTemperature sensors(&oneWire);
void setup(void)
{
 // start serial port
 Serial.begin(9600);
 Serial.println("Dallas Temperature IC Control Library Demo");

 Serial.println("This is SN002C Program by P&D Analytics");
  // Start up the library
 sensors.begin();
}
void loop(void)
{
 // call sensors.requestTemperatures() on sensors on Bus
 Serial.print(" Requesting temperatures...");
 sensors.requestTemperatures();
Serial.println("DONE");
Serial.print("Temperature is: ");
 Serial.print(sensors.getTempCByIndex(0)); // Why "byIndex"?
   // You can have more than one DS18B20 on the same bus.
   // 0 refers to the first IC on the wire
   delay(1000);
}
```

Getting it up and running

First, label the Arduino SN002C and load the code via the IDE to the Arduino UNO.

The code consists of three elements.

1. The first block of code loads the one wire and temperature sensor libraries, and sets up which port the sensor is connected to as well as the communication protocols between the computer and Arduino.

2. The next block of code sets up the baud rate, starts running the library, and begins the process of looking for data on the data bus.

3. The final section of the code requests temperature data and sends it over the serial port to the computer. It then repeats the process of gathering data.

Example data from the serial port.

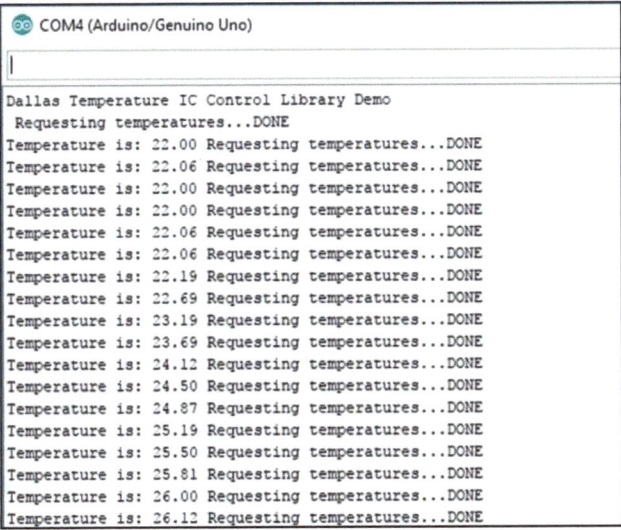

Figure 38: Serial Port Data for SN002B One Wire Temperature Sensor (°C)

How to use the data.

One use for this water proof one wire sensor is to gain insight related to *design* as a potential cause for the leaks (See Figure 39). This system could be placed in an enclosure in a hot, humid environment to gather data on how hot the sealed enclosure reaches in the actual environment. Also gather information to estimate the temperature changes between day and night.

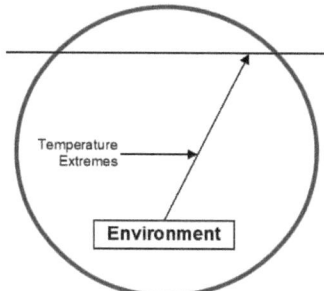

Figure 39: Section of Example Fishbone

Could the enclosure heating up during the day cause dimensional changes in the seal area that result in a leakage path?

This question was posed in the example problem, and an assumed thermal growth of the Lexan cover related to the aluminum housing from day to night temperature difference was calculated on page 21 in Table 1. The calculations assumed a temperature difference of 60 °F between day and night. It is assumed, given the Casting tolerance of 0.008 inches and the 0.007 calculated thermal growth in Table 1, that the compression of the seal would be lost. This might be especially true at the corners if the radius around the seal surfaces between the Lexan housing and the Aluminum enclosure don't match.

Based on this, it would be very beneficial to determine if the temperature difference is this great.

So a test was run using a simulated enclosure and using this water proof temperature sensor to determine a reasonable estimate for night to day temperature change.

Figure 40: Temperature change during the day

Estimated pressure increase

The temperature change as measured ranged from 78.8 to 120.2 °F. This is only 41.4 °F which is less than the 60 °F change assumed. Recalculating the thermal growth using a 41 °F results in a change 0.005 inches. Which would most likely not result in the seal loosing compression.

The next step is use to the temperature change to estimate the pressure increase based on a variation of the *Ideal Gas Law* that states if a volume is sealed the pressure and temperature for one state is proportional to the temperature and pressure of another state.

Or: $P_2 = P_1 \times (T_2 / T_1)$

Table 2: Data from Environment Temperature Test

p1 (kPa)	t1 (F)	t1 (C)	t2 (F)	t2 (C)	p1 (torr)	t1 (K)	t2 (K)	p2 (torr)	p2 (psi)	p2 (kPa)
101.35	75.2	24	78.8	26	760.19	297.15	299.15	765.31	14.80	102.03
101.35	75.2	24	86	30	760.19	297.15	303.15	775.54	15.00	103.40
101.35	75.2	24	96.8	36	760.19	297.15	309.15	790.89	15.29	105.44
101.35	75.2	24	98.6	37	760.19	297.15	310.15	793.45	15.34	105.78
101.35	75.2	24	100.4	38	760.19	297.15	311.15	796.01	15.39	106.13
101.35	75.2	24	102.2	39	760.19	297.15	312.15	798.56	15.44	106.47
101.35	75.2	24	105.8	41	760.19	297.15	314.15	803.68	15.54	107.15
101.35	75.2	24	107.6	42	760.19	297.15	315.15	806.24	15.59	107.49
101.35	75.2	24	111.2	44	760.19	297.15	317.15	811.36	15.69	108.17
101.35	75.2	24	113	45	760.19	297.15	318.15	813.91	15.74	108.51
101.35	75.2	24	116.6	47	760.19	297.15	320.15	819.03	15.84	109.20
101.35	75.2	24	118.4	48	760.19	297.15	321.15	821.59	15.89	109.54
101.35	75.2	24	120.2	49	760.19	297.15	322.15	824.15	15.94	109.88
101.35	75.2	24	120.2	49	760.19	297.15	322.15	824.15	15.94	109.88

The calculations above highlight the amount of pressure increase of approximately 1 psi that could be seen in the enclosure's environment. If the seal is properly seated based on the dimensions and clamping force these environmental conditions and resulting pressure increase should not cause a leak. Typically, there is quite a bit of O-ring compression in these types of seals, aspects of this type of seal are described in the Parker O-ring web handbook under triangular static seals:

https://www.parker.com/literature/ORD%205700%20Parker_O-Ring_Handbook.pdf

Therefore, it appears that if the dimensions and compression are within reasonable limits than the environmental and pressure increase for the operational temperature range is most likely not a cause for the leaks.

7. Measuring Light Intensity and Temperature

This chapter highlights methods to measure light and the combination of light and temperature. This can provide additional information to answer the question posed regarding high temperature causing the paint to run as shown in the repeated Figure 41 below. For this case the cause of the high temperature is light from a nearby window in the paint room. This Arduino system uses a unique sensor that measures the relative intensity of light at various wavelengths in the electromagnetic spectrum. The investigator can determine if the overall light intensity is sufficient energy to raise the temperature to a point where the paint runs into the seal area.

```
        ┌─────────┐
        │ Painting│
┌───────┴─────────┴───────┐
│ Hot                     │
│ Temperature             │
│ causing paint           │
│ to run into             │
│ seal area.              │
└─────────────────────────┘
```

Figure 41: Painting Step and Temperature Drying Question

There are two projects in this chapter. The first system uses a unique Adafruit light sensor TSL 2561, and it is directly connected to the Arduino which sends the data over the serial port to the computer. The second system combines the light sensor and a TMP 36 temperature sensor to aid in the determination of the significance of the relationship between light entering a room and the temperature increase that results.

Project 3: Arduino UNO SN003
Light sensor

Description/Goal: This system uses a unique sensor to measure light intensity. It is capable of separating different light frequencies, particularly in the infrared and visible light segments of the electromagnetic spectrum. These two segments contain most of the energy transmitted by light. The investigator will also learn to use a breakout board and measure incident light energy to determine heat transferred.

Challenges: The TSL 2561 light sensor code used in this book is relatively straight forward but it requires code that configures the sensor. The authors found the code below works well and it is important to determine which level of gain is needed for the project. Then use that gain consistently for useful comparisons. There is a lot of technical information pertaining to the sensor and how it works available but it is confusing pertaining to the method used to calculate the Lux output from the sensor. See the web site below for further information. However, the output from this sensor provides excellent data for comparison purposes.

https://learn.adafruit.com/tsl2561/overview?view=all

Hardware needed:

- Arduino Uno
- Computer
- TSL 2561 Light Sensor (The authors soldered a small header on the sensor board so it plugs right into a proto-board, see figure below)
-

The TSL 2561 is unique in it uses two photodiodes to measure the infrared and visible light. One photodiode is sensitive to infrared only and the other is sensitive to both visible light and infrared radiation, This incredibly low cost device then compares and integrates those measurements to calculate Lux (which is defined as Lumens/square meter) and outputs values for visible and infrared light.

The authors researched several specification sheets and websites for the units for visible and infrared light output from the sensor and could not determine the actual units. It is believed they are in Lumens but that may not be correct. However, the real use of this device is comparing two relative lighting situations and therefore the actual units are not critical.

Figure 42: SN003 Light Sensor

Figure 43: Light Sensor Header Installed

How to build the system

Assemble the sensor and Arduino per Figure 44 below.

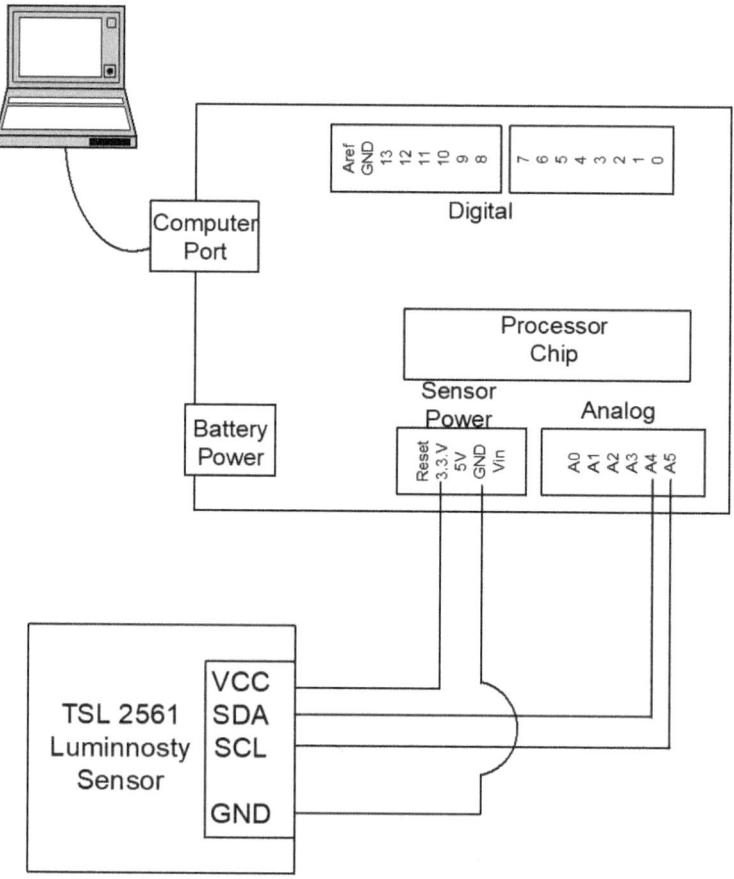

Figure 44: Light Sensor Schematic

This site contains a lot of very useful information, and it is where the basic elements for the code for the light sensor was obtained:

https://learn.adafruit.com/tsl2561/overview?view=all

SN003 Code is loaded using the IDE after the libraries Wire.h and TSL2561.h are loaded.

```
//SN003_Aug_26_LuxOnly

#include <Wire.h>
#include "TSL2561.h"

TSL2561 tsl(TSL2561_ADDR_FLOAT);

//Set up loop check for sensor
void setup(void) {
  Serial.begin(9600);

  Serial.println("This is program SN003_LuxOnly by P&D Analytics");

  if(tsl.begin()){
    Serial.println("Found Sensor");
  } else {
    Serial.println("No Sensor?");
    while(1);
  }

//Set gain and integration time

//Using zero gain for bright light
  tsl.setGain(TSL2561_GAIN_0X);

// 13MS for speed
  tsl.setTiming(TSL2561_INTEGRATIONTIME_13MS);
}

//Send data to serial port
void loop() {
  uint16_t x = tsl.getLuminosity(TSL2561_VISIBLE);

  Serial.print("The raw value for visible from sensor is: ");
```

```
    Serial.println(x, DEC);

    uint32_t lum = tsl.getFullLuminosity();
    uint16_t ir, full;

    ir = lum >> 16;

    full = lum & 0xFFFF;
    Serial.print("IR: ");
    Serial.print(ir);
    Serial.print("\t\t");
    Serial.print("Full: ");
    Serial.print(full);
    Serial.print("\t");
    Serial.print("Calculate Visible: ");
    Serial.print(full - ir);
    Serial.print("\t");

    Serial.print("Lux: ");
    Serial.println(tsl.calculateLux(full,ir));


    delay(10000);

}
```

Getting it up and running

The first step is to label the Arduino UNO SN003. The next step is to upload the code using the IDE.

The code consists of three primary sections.

1. The first lines of code load the libraries and initiate the light sensor.

2. The next section sets up the light sensor parameters. Note: The code above has a zero setting on gain which is for bright light.

3. At this point the program starts the process of monitoring, capturing, and sending the light sensor data over the serial port.

How to use the data

These data can be used to measure incident solar radiation from a window and help resolve the question regarding the heat transferred into painting process from our example root cause analysis.

The figure below shows light levels late in the afternoon when the room is warm.

This data can be compared to either a cloudy day or time when the room is relatively cool.

Then mitigation factors such as reflective film can be applied to the window and a comparison made to see if it reduces the amount of infrared or visible light entering the room. Also determine if there is a corresponding drop in temperature. The investigator must keep in mind that the source of heat is not just the light entering the room but the ambient temperature outside which increases significantly during the day. It is important to identify true causes and understand that in many cases there is more than one cause.

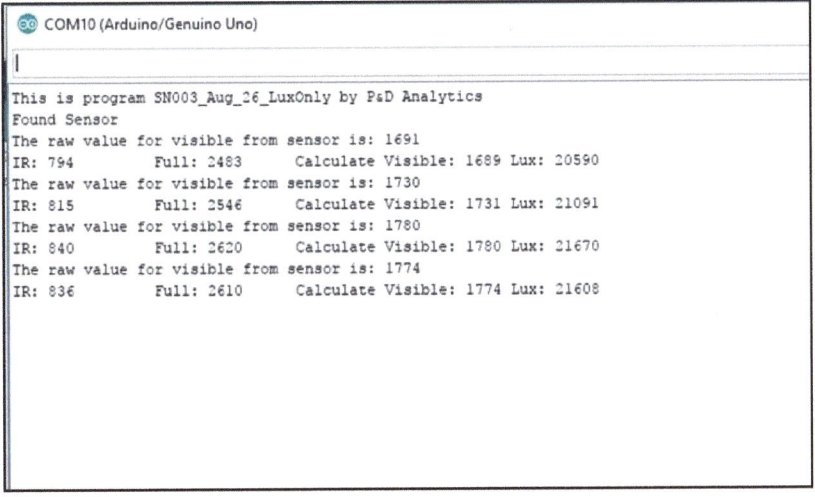

Figure 45: Afternoon Paint Room Light Sensor Data

The table below compares light intensity between morning and afternoon. The next step could be measuring it again after reflective film is installed to determine if that reduces the light entering the room.

Table 3: Average Visible and Infrared Light Entering Paint Rom

Morning or Afternoon	Average Visible	Average Infrared	Average Lux
Morning	14	8	163
Afternoon	1744	821	21240

The results above not surprisingly show that a significantly higher amount of light enters the room in the afternoon.

The next project which combines the light sensor with a temperature sensor can provide additional key data to aid the root cause team to determine if the light entering the room is really raising the temperature significantly.

Project 3A: Arduino SN003A

This system combines the previous project and the SN002 temperature sensor project. Using the TMP 36 sensor and the TSL 2561 connected to an Arduino, will help assess cause and effect related to sunlight entering the paint room and causing the temperature to rise.

Description/Goal: The goal of this project is to develop a system that measures and correlates the amount of light with the temperature. You will also learn about combining multiple sensors.

Challenges: The authors found the code for the TSL 2561 sensor below works well. It is important to determine which level of gain is needed for the project. Then use that gain consistently for useful comparisons. There is a lot of technical information pertaining to the sensor and how it works available but it is confusing pertaining how Lux is calculated. The real benefit of this sensor is for comparison purposes. Therefore the units are not that critical. See the web site below for further information.

https://learn.adafruit.com/tsl2561/overview?view=all

Also see the description of how the sensor works on page 84 in this book.

Hardware Needed:
- Computer
- Arduino
- Assembled Adafruit TSL 2561 Light Sensor Board
- TMP 36 Temperature Sensor

Figure 46: SN003A Light and Temperature Sensors

How to build the system

Build the system per the schematic below.

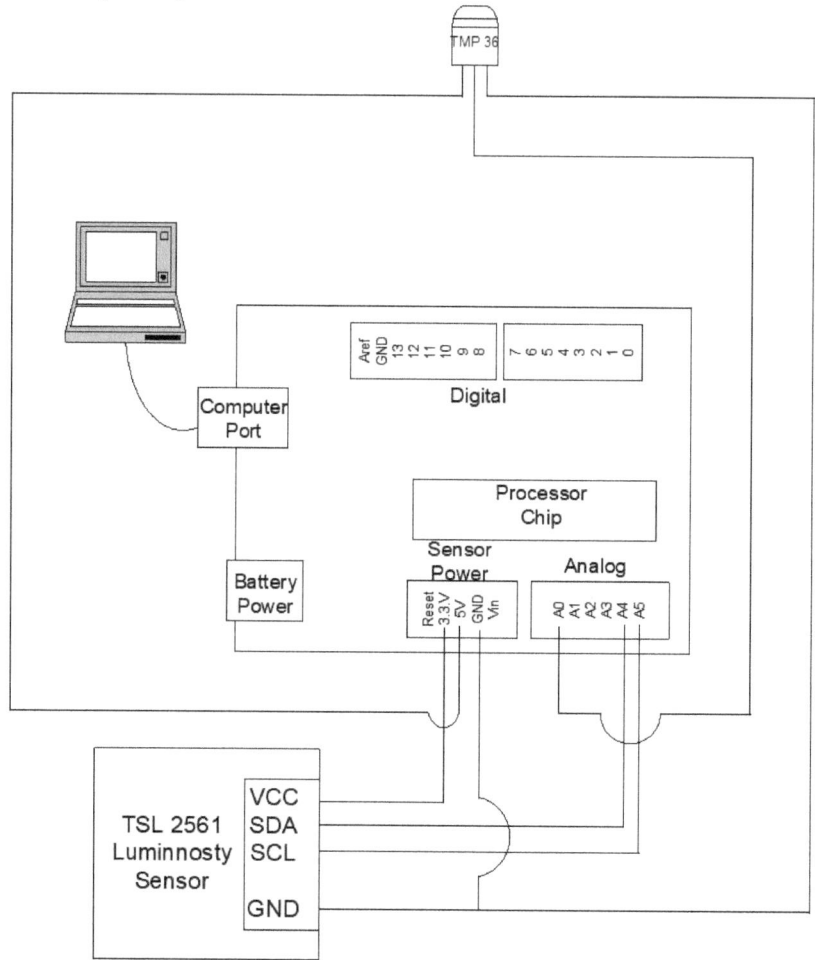

Figure 47: SN003A Light and Temperature

SN003A Code has TMP 36 sensor added from previous projects. This project needs the Wire.h and TSL2561.h libraries.

```
//SN003A_July_30_TempLux

#include <Wire.h>
#include "TSL2561.h"

TSL2561 tsl(TSL2561_ADDR_FLOAT);

//Set up Temp variables
const int analogIn = A0;

int RawValue = 0;
double Voltage = 0;
double tempC = 0;
double tempF = 0;

//Set up loop check for sensor
void setup(void) {
  Serial.begin(9600);

  Serial.println("Program SN003A_TempLux by P&D Analytics");

  if(tsl.begin()){
    Serial.println("Found Sensor");
  } else {
    Serial.println("No Sensor?");
    while(1);
  }

//Set up gain and integration time

//Gain is zero for bright light
  tsl.setGain(TSL2561_GAIN_0X);

//13MS for speed
  tsl.setTiming(TSL2561_INTEGRATIONTIME_13MS);
}

//Send data to serial port
void loop() {
  uint16_t x = tsl.getLuminosity(TSL2561_VISIBLE);
```

```
Serial.print("The raw value for visible from sensor is: ");
Serial.println(x, DEC);

uint32_t lum = tsl.getFullLuminosity();
uint16_t ir, full;

ir = lum >> 16;

full = lum & 0xFFFF;
Serial.print("IR: ");
Serial.print(ir);
Serial.print("\t\t");
Serial.print("Full: ");
Serial.print(full);
Serial.print("\t");
Serial.print("Visible: ");
Serial.print(full - ir);
Serial.print("\t");

Serial.print("Lux: ");
Serial.println(tsl.calculateLux(full,ir));

RawValue = analogRead(analogIn);
Voltage = (RawValue / 1023.0) * 5000;
tempC = (Voltage - 500) * 0.1;
tempF = (tempC * 1.8) + 32;

Serial.print("The raw milli volts is: ");
Serial.print(Voltage, 0);
Serial.print(", Raw Temp Value is: ");
Serial.print(RawValue);
Serial.print("; Temperature in C = ");
Serial.print(tempC,1);
Serial.print("; Temperature in F = ");
Serial.println(tempF,1);


delay(10000);

}
```

Getting it up and running

The first step is to label the Arduino UNO as SN003A. Upload the code using the IDE.

The program consists of five primary elements:

1. The first lines of code load the libraries and initiates the light sensor.
2. The next section sets up the temperature sensor parameters.
3. The next section of code sets the gain for the light sensor.
4. At this point the program starts the process of monitoring, capturing, and sending the light and temperature sensor data over the serial port to the computer.
5. The last set of code sends the light and temperature data over the serial port to the computer.

How to use the light sensor data

The data captured by the light sensor can provide relative information. Compare it from one day to the next to see changes. Select a standard time interval when the sun is shining in the window, and then repeat the measurement at about the same time of day for several days to gain an insight. Another approach would be to gather data on a cloudy day and a sunny day and observe how the clouds affect the intensity of light coming in the window.

The table below compares some of these measurements.

Table 4: Visible, Infrared Light, and Temperature Entering Paint Room

Morning or Afternoon	Average Visible	Average Infrared	Average Lux	Average Temperature (°F)
Morning	7	14	159	73
Afternoon	828	1716	20727	87

There is some evidence that the cause of the leak is from paint running into the seal due to the high temperature. One potential cause of the high temperature is the light through the window during the late afternoon.

However the ambient outside temperature of 91 °F may be a bigger factor in the temperature increase in the paint room. Therefore the light coming in the window is a factor but may not be the primary aspect.

For further information related to heat transfer and the factors that affect and influence how heat flows, see Appendix G.

8. Measuring Pressure Change

In the example problem there was a concern that the pressure test might cause the O-ring to shift and not seal. This system can be used to answer the last question the root cause analysis team posed, could the pressure test displace the seal. The system uses a small pressure sensor and an Arduino connected to the computer.

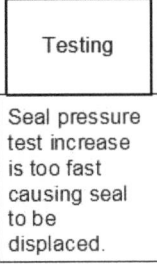

Figure 48: Pressure Testing Question

Project 4: Arduino SN004
Pressure measurement change during seal test

Description/Goal: This system is a little complex because this type of sensor requires the addition of an amplifier to boost the signal to a level that can be registered by the Arduino. The investigator will learn how to connect the Arduino to external circuits that condition the signals from pressure sensors. Additionally, the investigator will learn about the Wheatstone bridge which is embedded in the pressure sensor.

Challenges: The challenge for this project is building the interface circuit using the op-amp device. The op-amp or operational amplifier is a very useful device that can boost signals out of the noise to a level that they can be easily measured. One other challenge is the calibration of the pressure sensor using the Arduino. The website below has some useful information related to this pressure sensor:

https://www.digikey.com/products/en?mpart=26PCBFA6D&v=480

Hardware needed:
- Arduino Uno
- Computer
- Honeywell Pressure Sensor 26PCBFA6D (Low cost sensor)
- INA125P Operational Amplifier

How does a Wheatstone Bridge in the pressure sensor work?

The Wheatstone Bridge (See Figure below) is a common technique used to compare and measure small changes in a resistance. Often load cells and pressure transducers have a Wheatstone bridge inside. One or more resistors are on the membrane that will flex based on the load applied in a load cell or the pressure in a pressure sensor. A voltage is applied across the bridge and if there is a difference in each leg of the resistance then a small voltage drop can be seen between pin 3 and 4. This is small, so it needs to be amplified to a level that can be measured.

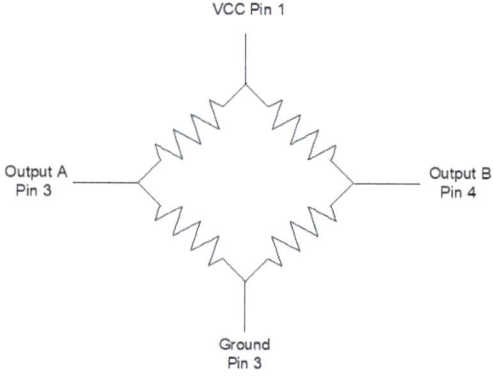

Figure 49: Wheatstone Bridge and Pressure Sensor

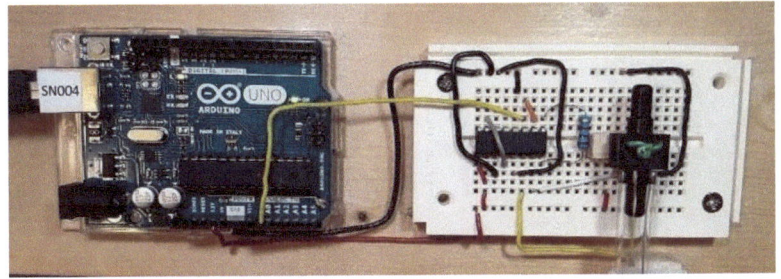

Figure 50: Assembled SN004 Pressure Measurement Sensor

How to build the system

Build it per the schematic below.

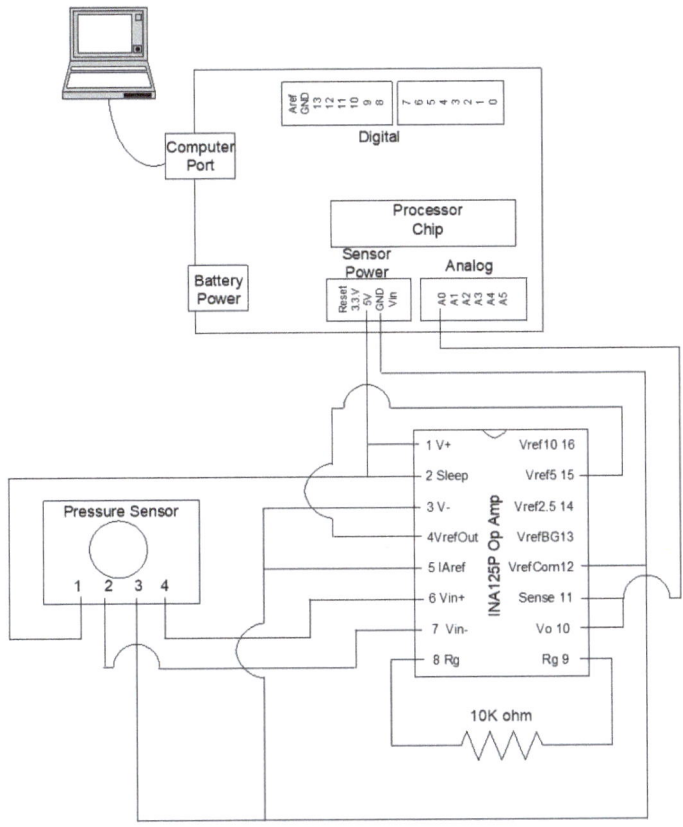

Figure 51: SN004 Pressure Sensor Schematic

99

Figure 52: Attach Pressure Sensor to Proto-Board

Code needed

The website below provides the information related to building the interface circuitry to amplify the signal from a load cell. Which operates in a similar manner to the pressure sensor in this project.

https://edg.uchicago.edu/tutorials/load_cell/

The SN004 code is modified in a few areas from the code on this site.

SN004 code needs to be loaded using the IDE.

```
// SN004_Pressure_July_1
// Orignal code is from U Chicago
// Calibrate by reading value at zero psi and 10 psi.
// Put those values here.
float aReading = 14.0;
float aLoad = 0; // psi
float bReading = 680.0;
float bLoad = 10; // psi

long time = 0;
int interval = 5000; // Take a reading every 5 sec

void setup() {
  Serial.begin(9600);

  Serial.println("This is SN004 Program by P&D Analytics");
}

void loop() {
  float newReading = analogRead(0);

  // Calculate load based on A and B readings above
  float load = ((bLoad - aLoad)/(bReading - aReading)) * (newReading - aReading) + aLoad;

  // millis returns the number of milliseconds since start of program
  if(millis() > time + interval) {
    Serial.print("Reading: ");
    Serial.print(newReading,1); // 1 decimal place
    Serial.print("  psi: ");
    Serial.println(load,3);  // 3 decimal places
    time = millis();
  }
}
```

Note this is all on one line of code not two: The length of this line pushed the code into two lines, however it is all one line in the IDE.

float load = ((bLoad - aLoad)/(bReading - aReading)) * (newReading - aReading) + aLoad;

Getting it up and running

The first step is to label the Arduino UNO SN004. Next, load the code using the IDE.

The program consists of three primary elements:

1. The first part sets up the data elements and time between readings.

2. The next section sets up the baud rate and calculates the pressure.

3. The last set of code sends the pressure reading over the serial port to the computer.

Next step need to calibrate the Arduino output by measuring the pressure by taking the readings for zero psi and 10 psi. The authors used 10 psi as the sensor is rated to 20 psi over pressure. Then modify the corresponding values in the code of a and b reading along with a and b load. Then the system is ready to use.

Figure 53: Testing Pressure Rate for Enclosure

How to use the data

The following graph shows the calibration the authors created for the sensor. Using the readings of 14 equal to 0 psi and 680 equal to 10 psi, the graph below was also drawn to show an assumed linear relationship.

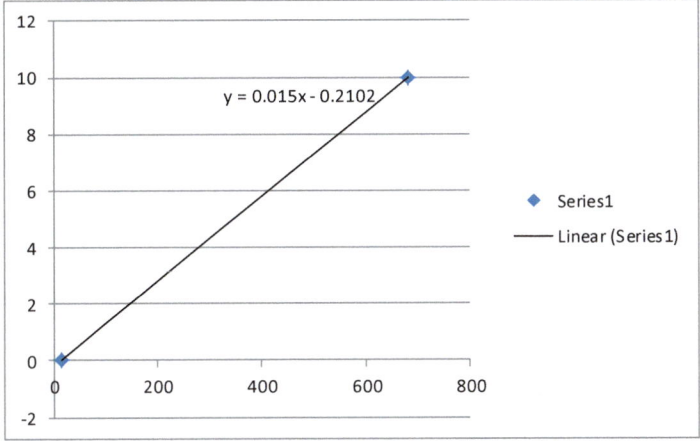

Figure 54: Calibration Curve for Pressure Sensor

With this calibration input into the code the following reading was taken on two separate occasions for the enclosure leak test. See the graph below for the results of the test. As always the pressure spikes up and then settles out at a lower value. In this case the seal did not appear to be displaced so this indicates that this is not a valid cause.

Slow Pressure Increase		Fast Pressure Increase	
Time (sec)	Pressure Slow	Time (sec)	Pressure Fast
0	0.00	0	0.00
5	8.11	5	9.99
10	9.97	10	9.27
15	9.28	15	8.53
20	8.61	20	8.49

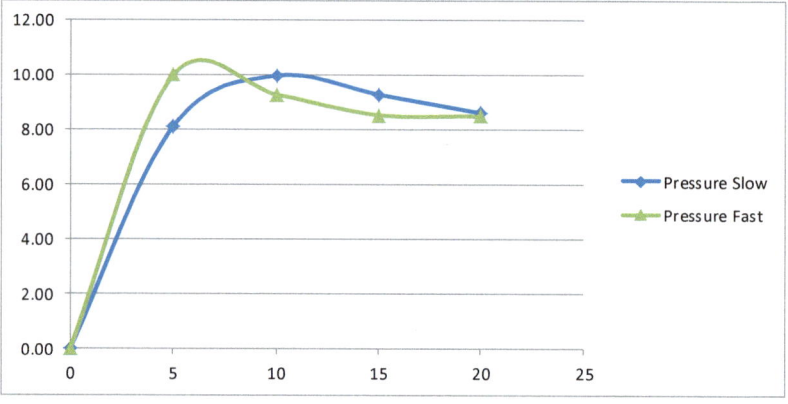

Figure 55: Pressure Test Results

It is important to ensure the system is leak tight prior to connecting to the enclosure, if there is a small leak in the tubing it can mask what is happening.

9. Root Cause Analysis of Measured Data

This section shows how to wrap up a root cause investigation. Two aspects are key to this part of the analysis. The first is a list of the suspected causes. If an item is not on the list, then it is not being considered as a cause or causal factor. The list also conveys what the team has investigated.

The second aspect is using a method that weights the evidence based on rationale. For example, a three- or five-level weighting scheme can be used based on evidence or engineering judgement to rank the causal factors. Table 5 below is a three level system of strong, medium, and weak evidence. One hopes the data either excludes or justifies each cause. However, in many cases the evidence may not be strong enough to provide rationale one way or another. This method communicates this uncertainty in the analysis and helps focus on areas to try and gather more evidence. It is a very good idea to start on this evidence table early in the root cause effort as it can guide the team.

Table 5: Evidence Weighting

Cause Ranking	Weighting of Evidence	Description and Evidence
1	Strong Evidence	**Expedited Subassembly and Assembly operations:** See Figures 21 (pg 45) and 25 (pg 53) which show definite trends showing a decrease in the time to complete these tasks.
2	Medium Evidence	**Temperature increase causing paint to run into seal area:** See Figures 31 (pg 62) and Table 4 (pg 95) which show evidence of increases in temperature in the paint room, however because of the ambient temperature it may not be significant.
3	Weak Evidence	**Design of Seal/Temperature Extremes:** See Table 1 (pg 21) and Table 2 (pg 81) which show that the design should be adequate to handle the environmental conditions.
4	Weak Evidence	**Fast Pressure Test displacing seal:** See Figure 56 (pg 104) which shows that there does not appear to be a difference between the seal performance between the fast and slow test.

The evidence above points to expedited assembly as the primary root cause. It is very important to understand why these steps are being expedited. Is it a real requirement that is driving rushing these steps or is it some other aspect? It is important to find the factors that influence people's behavior and work on changing them in a positive way. Stay away from the blame game, focus on improvements that help everyone, both management and the work force.

10. Enclosures

These last projects highlight ways to protect an Arduino data system that could be damaged by activity around it. Installing the Arduino in an enclosure can protect it. Many Arduino enclosures are available, but they usually are small, and it is often difficult to fit connections into them. An example of this is shown below. A set of terminal strips were soldered into a PC board that was cut to fit in an opening in the enclosure. A small plate was then cut to fit around the terminal strips with two holes that screws went through into the housing. This held the terminal strips in place. This looks reasonably professional, but was relatively difficult to fabricate. The terminal strip was small and could not accept large wires.

Figure 56: Arduino Specific Enclosure

The authors suggest using a generic enclosure where there is plenty of room for the connection terminal strips. Using a generic enclosure means that it must be modified so that the Arduino can be mounted, and access must be provided for the power or serial port connections too.

The following projects provide some examples of how to use off-the-shelf standard items as an enclosure for any of these Arduino projects.

Description: This section describes three enclosure methods and some tips and tricks to consider. The three methods are:

1. *Integrated Arduino on the cover plate* (See figure below) This enclosure style uses a standard project box and the Arduino is mounted to the cover plate. The terminal strip is mounted on the outside of the cover plate and the wires pass through from the Arduino. The side has to have a cutout for access to the SD card, the computer port and the power plug.

Figure 57: Arduino Mounted to the Cover

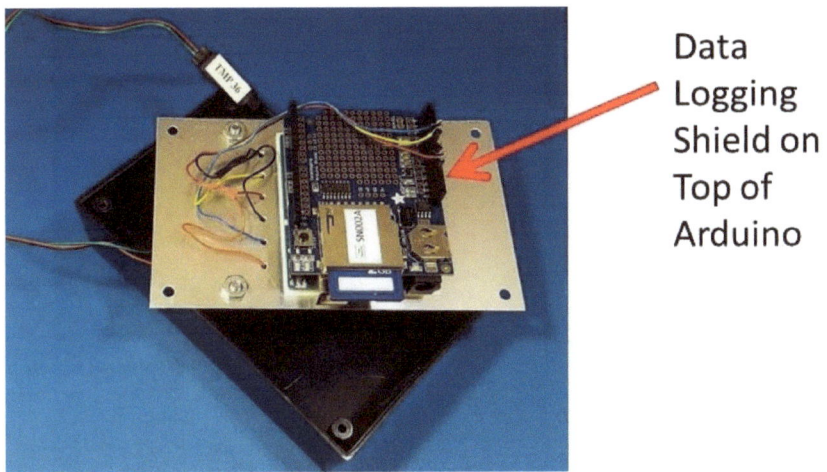

Figure 58: Internal Arrangement of Arduino Mounted to the Cover Plate

Figure 59: Side View of Connections and SD Card

2. *Arduino mounted on a board with an inexpensive cover mounted over it* (See figure below). This is a narrow wiring breakout box for a wall switch. Since all of the Arduino and other components are mounted on a board, the only modifications were two holes cut out for the serial connection and the power plug. The wires to the temperature sensor were passed under a small gap between the mounting tabs.

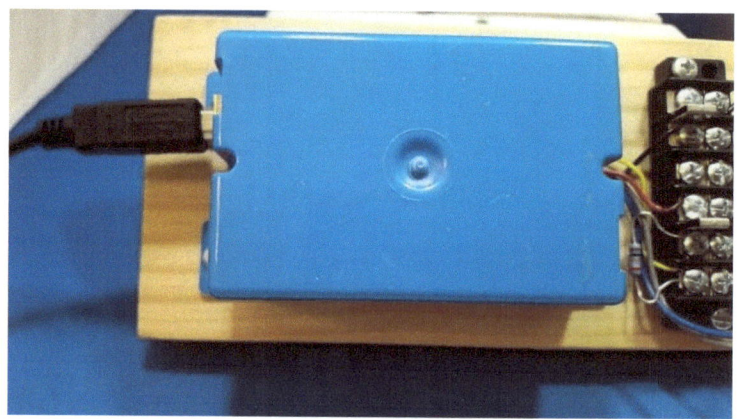

Figure 60: Simple Enclosure

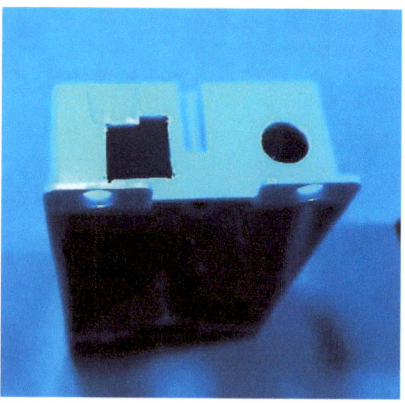

Figure 61: Serial Port and Power Holes in Box

3. *Water proof Pelican® case* (Figure 62). This next configuration can be used if the environment is very humid or wet. The cables for the sensor will need to pass through a hole with a grommet and then the liner will be pushed up against it surrounding the wires. Having these holes defeats the ability of the Pelican® case to seal out water while submerged. It will keep water that is splashed on it from entering the case. See the figures below for more details. The first hole for the serial connector will need to start as a drilled hole and then file it into a square shape so that the connecter can pass through. To assemble this seal need to install the grommet over the wire, then pass the connector through the hole. Finally, install the grommet in the hole. The next step is cut the liner so it can fold down and has a hole for the wire to pass through.

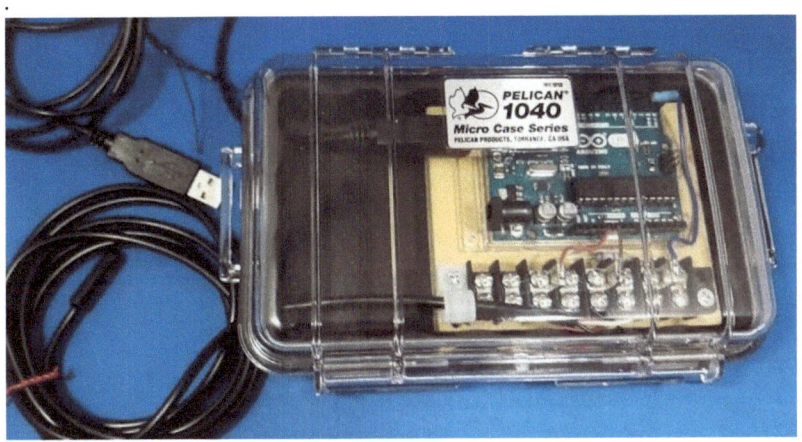

Figure 62: Water Proof Pelican® case

The next figure shows how the flaps of the liner are cut so they can fold back. Also cut out a small hole in the flap. So that after the wire passes then the flap is pushed back into place to provide a secondary water block.

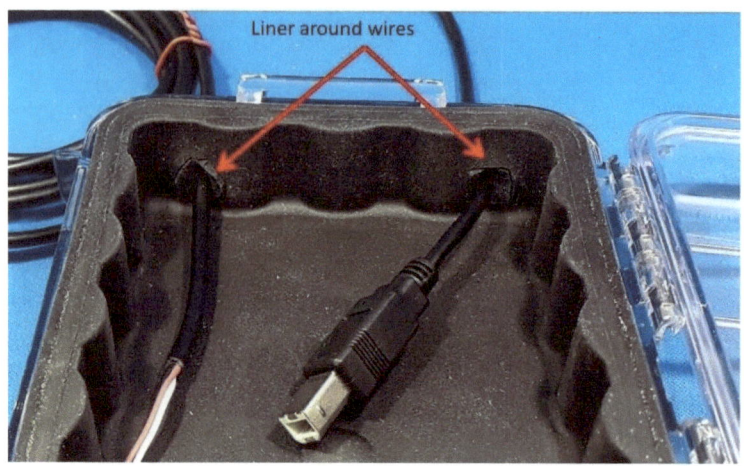

Figure 63: Liner Flaps that Wrap Around the Wires

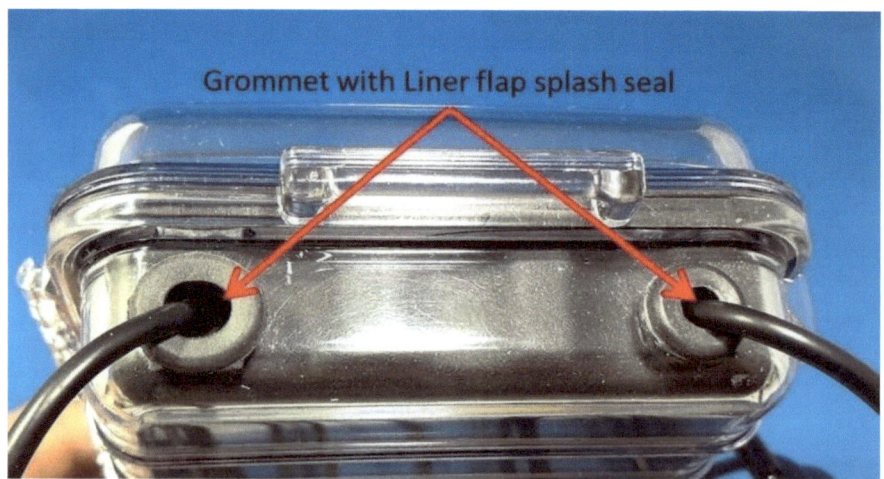

Figure 64: Wires through the Grommet and Liner Splash Seals

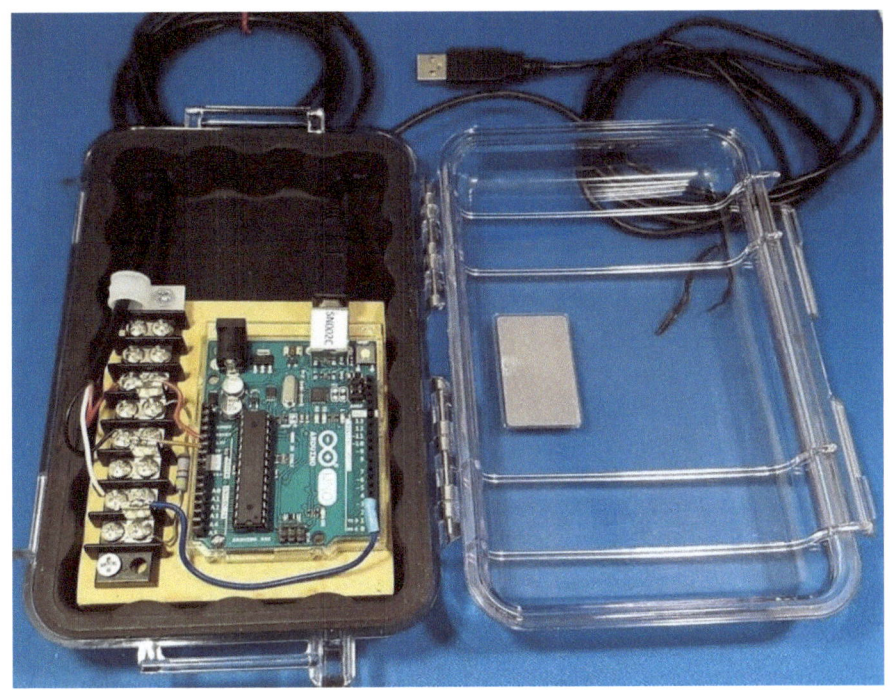

Figure 65: Pelican® Case and Arduino

Another option rather than using a grommet is the use of a compression fitting. This can provide a better seal.

Figure 66: Compression Fitting that Seals Wires

P & D Analytics Incorporated

Conclusion

This book provides an example problem and then works through analyzing some potential causes. Using the root cause process steps ensures that the team will develop a valid cause analysis for the problem. There are aspects of the example root cause analysis of the seal problem described in Chapter 2 that are very similar to three failures that P & D Analytics Inc. personnel observed in the past. The potential causes all played a role in the real-life observed failures. The authors also observed that seals at the interfaces of subassemblies could fail when combined with adverse environmental conditions.

The real world has many challenges with obtaining data and insight into problems. The Arduino and the large number of sensors available for it can help answer many questions that previously might have been considered unanswerable, or at least, very expensive and time consuming to answer. Quickly solving these issues can reduce expensive recalls and product redesigns. Arriving at real root cause solutions can significantly impact in a positive way the company's bottom line.

The authors hope this book provides the guidance and tools to aid root cause teams on their quest to find causes and develop solutions to their problems.

Some other questions that an Arduino-based system using existing sensor technology may aid in answering are:

- Identifying electrical power-hungry devices with an external power monitor.
- Measuring temperature stratification in a tall room.
- Measuring the temperature and humidity of an area.
- Determining solar incidence on a roof to aid in solar energy decisions.
- Monitoring external pipe temperatures from a water heater and assessing insulation effectiveness.
- Measuring ambient radiation levels.
- Measuring the acceleration of an item as it is transported from location to another.
- Measuring the distance or speed an object moves using an ultrasonic motion sensor.

These are other opportunities for a root cause teams to make a difference.

P & D Analytics Incorporated

Appendix

A. Sources of Hardware

B. Other Sensors that Could be Substituted

C. Soldering Safety Tips

D. Encapsulating and Calibrating the TMP 36 Temperature Sensors

E. Data Logging Shield Assembly

F. Measurement Uncertainty Analysis

G. Heat Transfer Equations

H. Material Properties, Casting Tolerance, and O-Ring Compression Information

P & D Analytics Incorporated

A. Sources for Hardware

This is a list of some very good hardware suppliers. It is not comprehensive, as there are many other great sources for parts.

Adafruit
https://www.adafruit.com/

Arduino
https://www.arduino.cc/

Bud Enclosures
http://www.budind.com/

Digikey
https://www.digikey.com/

Fry's Electronics
http://www.frys.com/

Mouser Electronics
http://www.mouser.com/

Online Electronics Distributor
onlinecomponents.com

Pelican® Cases
http://www.pelican.com/

Sparkfun
https://www.sparkfun.com/

P & D Analytics Incorporated

B. Other Sensors that Could be Substituted

An alternate for the Sparkfun Real Time Clock is from Adafruit

https://www.adafruit.com/product/3296

An alternate for the TMP 36 temperature sensor is the LM 35 sensor.

http://www.ti.com/lit/ds/symlink/lm35.pdf

An alternate for the Adafruit Light sensor is the Sparkfun TSL 2561 light sensor breakout board.

https://www.sparkfun.com/products/12055

There are many alternates for the 26PCBFA6D pressure sensor, including many different Honeywell pressure sensors. The 26PCBFA6D was selected for its relatively low cost and that it was readily available.

An alternate for the Op Amp used in SN004 project is the INA 122 which is a smaller format with only 8 inputs.

P & D Analytics Incorporated

C. Soldering Safety

Soldering can be an enjoyable experience, but occasionally a difficult soldered connection can also be a frustrating experience. These tips will help make it a safe experience.

- First make sure you understand the basics of soldering. If unsure, either take a basic class in electronics or find someone who can demonstrate the basics.

- Practice on scrap wires.

- Use fixtures, objects, or tape to hold the pieces in the proper configuration.

- Avoid breathing the fumes from the soldering iron.

- Always remember the soldering iron tip and recently heated soldered connections are very hot.

- Always remember to wear safety glasses to protect your eyes.

- Remove flammable objects from the surroundings.

P & D Analytics Incorporated

D. Encapsulating and Calibrating the TMP 36 Temperature Sensors

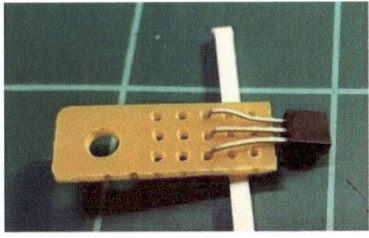
Step 1: Insert Sensor Wires

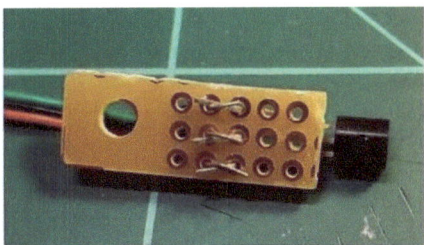
Step 2: Insert Cable Wires

Step 3: Attach Heat Sink

Step 4: Solder Wires

Step 5: Insert Cable Thru Hole

Step 6: Install Heat Shrink

Figure 67: Assembly of TMP 36 Sensor

The steps above show a method to encase the TMP36 or other small transistor sensors.

The sensor may need calibration and the best way to do that is to compare its reading to a known temperature measurement device at two different temperatures. Then adjust this line of code so that it matches the "standard" being used.

   Voltage = (RawValue / 1023.0) * 5000; // 5000 to get millivots.

P & D Analytics Incorporated

E. Data Logging Shield Assembly

The Adafruit Data Logger Shield is shown in the figure below. This shield plugs into the top of the Arduino board, and it can write the data on an SD memory card.

Figure 68: Assembled SD Card Data Logging Shield

This breakout board has an onboard, Real Time Clock which can synchronize with the computer time so when disconnected, it can list the time the data were taken. Listing the actual time can be very helpful in assessing issues by identifying time-of-day relationships and influences that may lead to identifying a cause.

The assembly of this Adafruit shield requires some soldering See the Adafruit website for instructions to assemble:

https://www.adafruit.com/product/1141

The instructions on the Adafruit website are good and require paging through each section. The next figure show soldering the headers in place. The authors did not use the headers supplied with the kit, but preferred the through headers which allow direct wire insertion connections through the shield to the Arduino below it.

Figure 69: Soldering the Headers to Data Logger Shield

F. Measurement Uncertainty Analysis

The process of measurement often includes many different pieces of equipment and environmental factors. Each one of these elements can add a small error and cause the measurement to be slightly off or uncertain. There is a methodology to address uncertainty and calculate the effects from each element. It is a relatively simple calculation and shows which item drives the magnitude of the total error. The formula for total uncertainty or the statistical tolerance is the square root of the sum of the squares of each tolerance (See equation below).

For example, in the DS18B20 temperature sensor system project SN002B, there is a resistor and a temperature sensor. In the specification there is a graph showing the variability ±0.05°C over the entire temperature range. Next the resistor utilized has red tolerance band which indicates ±0.02 change in resistance. For this analysis will assume that the resistor directly impacts the temperature measurement. Inserting these values into the equation below yields the system's total uncertainty.

Total Uncertainty = (Rtol² + Sensortol²)$^{1/2}$

DS18B20 System Total Uncertainty = ((0.02)² + (0.05)²)$^{1/2}$ = ± 0.053 °C

If this uncertainty is too large (though it probably should be adequate for most cases), then limiting the temperature range can improve it. The graph for the DS18B20 shows that if the temperature range is limited to 20 to 30 °C then the tolerance is more like ± 0.03.

Inserting this value into the equation yields these results.

DS18B20 System Total Uncertainty = ((0.02)² + (0.03)²)$^{1/2}$ = ±0.036 °C

Another item that is not considered is the sensitivity of the Arduino as it measures the inputs. For the DS1820 sensor this is not a factor, because we are reading data on the digital port and not an analog voltage.

However, this is not the case for the TMP 36 temperature sensor. In this case, as the Arduino is reading the voltage on an analog port, it may be introducing an error to the system. The way the program SN002 works is when it sets up the analog port, it converts the 1023 bit range to 5 volts. It then multiplies the reading by 5000 to convert it to millivolts.

https://www.arduino.cc/en/Reference/AnalogRead

The forum listed below discusses some potential issues related to the analog ports on the Arduino.

http://forum.arduino.cc/index.php?topic=159271.0

For our calculation purposes let's assume that the error introduced from the Arduino is ±0.04. We will assume that this impacts the temperature reading directly.

From the Sparkfun site information

https://www.sparkfun.com/products/10988

The TMP 36 sensor has a ±0.02 °C.

The combined Total Uncertainty = $((0.04)^2 + (0.02)^2)^{1/2}$ = ±0.045 °C

So as long either sensor is providing consistent data, they are both relatively accurate given the system configuration and impacts from other components.

G. Heat Transfer Equations

This information can aid in calculating impacts from heat migrating due to high temperature from conduction, convection, or radiation. It is important to understand what factors drive heat transfer, as this may provide ideas for eliminating or mitigating heat migration into areas that can cause problems.

For conduction heat transfer the following equation provides the investigator with a method for determining the effect of heat migrating through material.

Conduction heat transfer equations:

The equation is:

$$dq/dt = (k (T_h - T_c)/dx$$

Where:
  dq/dt = heat transfer (Joule/sec)

  k = Conduction heat transfer coefficient (Joule/sec-m- °C) Material dependent

  dA = Area heat is transferred across

  T_h = Temperature at hot location (°C)

  T_c = Temperature at cold location (°C)

  dx = material length or thickness (m)

Convection heat transfer equations:

This next equation shows the factors that can aid in the estimation of convection heat transfer. Convection is the term used when a fluid like air or water carries heat away or adds it to an object it flows over.

$$dq/dt = h\, dA\, (T_h - T_c)$$

Where:

dq/dt = heat transfer (Joule/sec)

h = surface convection heat transfer coefficient (Joule/sec-m² - °C) The value of this variable can become very complex to determine as it is dependent on the gas, the conditions on the surface, the temperature, humidity, velocity of flow across it, and other items. There are some methods to determine this value in heat transfer texts or online sources.

dA = area through which heat is transferred (m2)

T_h = Temperature at hot location (°C)
T_c = Temperature at cold location (°C)

Radiation heat transfer equations:

The term radiation as it is used in heat transfer is different than its normal use. It does not necessarily result from a nuclear reaction, but it does require a very hot object. The object is so hot (like the sun) that it actually gives off energy in the form of light, and the light is the medium that transfers the energy.

$$dq/dt = \varepsilon \sigma A (T_1^4 - T_2^4)$$

Where:

dq/dt = heat transfer (Joule/secs)

ε = emissivity is a factor of the material and indicates how much of the energy is emitted due to the internal temperature

σ = Stefan-Boltzmann constant = 5.67×10^{-8} W/m² – K⁴

A = area through which heat is transferred (m²)

T_1 = Temperature at hot location (K°)

T_2 = Temperature at cold (K°)

P & D Analytics Incorporated

H. Material Properties, Casting Tolerances, and O-Ring Compression Information

These sites provide information regarding thermal growth or expansion coefficients. It is interesting to note that plastics have larger growth rates than metals. Also, aluminum has a higher growth rate than steel.

Lexan or polycarbonate:

http://www.associatedplastics.com/forms/pc_lexan_9034.pdf

Aluminum and other materials:

http://www.engineeringtoolbox.com/linear-expansion-coefficients-d_95.html

This is a very good site that has casting tolerance for various plastics:

http://www.designinfosystem.com/index.php?option=com_content&view=article&id=55&Itemid=57

These resources provide a lot of data related to the design and application of O-rings.

https://www.parker.com/literature/ORD%205700%20Parker_O-Ring_Handbook.pdf

http://www.applerubber.com/src/pdf/section3-o-ring-basics.pdf

P & D Analytics Incorporated

Bibliography

1. "Adafruit Data Logger Shield" [Online]. https://learn.adafruit.com/adafruit-data-logger-shield [Accessed 1/26/2017].

2. "Adafruit TSL 12561 Light Sensor" [Online]. https://learn.adafruit.com/tsl2561/overview?view=all [Accessed 2/15/2017].

3. "Arduino" [Online]. https://www.arduino.cc/ [Accessed 6/1/2017].

4. Banzai, M., Getting Started With Arduino, Sebastopol: O'Reilly, 2011

5. Bonner, Michael, "The Importance of Controlling Coating Temperature in UV Applications" [Online]. http://www.radtech.org/proceedings/2014/papers/Measurement%20&%20Analysis/Bonner%20-%20The%20Importance%20of%20Controlling%20Coating%20Temperature%20in%20UV%20Application%20Processes.pdf [Accessed 5/29/2017].

6. Dimitrov, Konstantin, "Arduino Thermometer with DS18B20" [Online]. https://create.arduino.cc/projecthub/TheGadgetBoy/ds18b20-digital-temperature-sensor-and-arduino-9cc806 [Accessed 5/19/2017].

7. Henry's Bench, 'TMP 36 Temperature Sensor' [Online]. http://henrysbench.capnfatz.com/henrys-bench/arduino-temperature-measurements/simple-tmp-36-arduino-thermometer/ [Accessed 5/26/2017].

8. "Sparkfun Real Time Clock code" [Online]. http://bildr.org/2011/03/ds1307-arduino/ [Accessed 5/27/2017].

9. "Working with Load and Cell and Arduino" University of Chicago, [Online]. https://edg.uchicago.edu/tutorials/load_cell/ [Accessed 6/10/2017].

P & D Analytics Incorporated

About P & D Analytics Inc.

P & D Analytics Inc. is a small business located in Houston, Texas. The company is dedicated to helping others improve their root cause analysis skills. The principles at P & D Analytics Inc. (Paul Bradt and David Bradt) bring a wealth of experience in troubleshooting technical and process issues, along with knowledge of the Arduino system and some of the unique measurement aspects this versatile microcontroller provides.

If you have a significant issue or problem, the authors hope this book may provide some possible solutions to gain more insight into causes and influencing factors.

www.ingramcontent.com/pod-product-compliance
Lightning Source LLC
Chambersburg PA
CBHW040217220526
45473CB00001B/22